Nupur Goyal, Akansha Gupta, Mangey Ram (Eds.)

AI-Infused Mathematical Modelling

De Gruyter Series on the Applications of Mathematics in Engineering and Information Sciences

---

Edited by
Mangey Ram

Volume 27

AI-Infused Mathematical Modelling

Computational Intelligence and Predictive Analysis

Edited by
Nupur Goyal, Akansha Gupta and Mangey Ram

DE
G
DE GRUYTER

Editors

Dr Nupur Goyal
Department of Mathematics
Graphic Era (Deemed to be University)
Clement Town
Dehradun 248002
Uttarakhand
India
nupurgoyalgeu@gmail.com

Prof. (Dr) Mangey Ram
Department of Mathematics
Graphic Era (Deemed to be University)
Clement Town
Dehradun 248002
Uttarakhand
India
mangeyram@gmail.com

Dr Akansha Gupta
Department of Computer Science and Engineering
Graphic Era (Deemed to be University)
Clement Town
Dehradun 248002
Uttarakhand
India
akanksha3000@gmail.com

ISSN 2626-5427
ISBN 978-3-11-914601-2
ISBN 978-3-11-221199-1 (PDF)
ISBN 978-3-11-221247-9 (EPUB)
DOI https://doi.org/10.1515/9783112211991

Library of Congress Control Number: 2026938092

Bibliographic information published by the Deutsche Nationalbibliothek
The Deutsche Nationalbibliothek lists this publication in the Deutsche Nationalbibliografie; detailed bibliographic data are available on the Internet at https://dnb.dnb.de.

De Gruyter and Walter de Gruyter GmbH are part of De Gruyter Brill.
www.degruyterbrill.com

Questions about General Product Safety Regulation:
productsafety@degruyterbrill.com

Cover illustration: MF3d/E+/Getty Images
Printing: CPI books GmbH, Leck

Preface

This book explores the integration of artificial intelligence (AI) with mathematical modeling to solve contemporary problems. It covers AI-enhanced predictive models, optimization techniques, and practical applications across various fields, highlighting the transformative impact of AI on traditional mathematical approaches.

This book is essential for mathematicians, data scientists, researchers, students, and industry professionals, offering innovative solutions and guidance on integrating AI into mathematical practices with a focus on practical, ethical, and interdisciplinary approaches. Some studies with applications are discussed in the following chapters.

Chapter 1 highlights the applicability of numerical and computational modeling techniques, commonly used in physical and engineering systems, for analyzing complex, interaction-driven systems beyond traditional content domains. In this study, YouTube is used as a representative complex system, and a dynamic multistage modeling framework is proposed to analyze the interactions among three interdependent system variables, namely views, likes, and subscribers for a given video.

Chapter 2 charts the comprehensive journey of predictive modeling, from its foundational principles to its deployment in the real world and its future trajectory.

In Chapter 3, the present knowledge of glacier mass balance, observational monitoring methods, and prediction techniques – from process-based, physically robust models to new machine learning (ML) approaches – are summarized.

Chapter 4 discusses the evolving domain of the gig economy, focusing on how AI is integrated with it and how it affects both gig workers and businesses. It focuses on the current trends where the AI and other computational intelligence techniques are discussed enabling the reshaping of work dynamics.

Chapter 5 shows the applicability of a deep learning (DL) algorithm, the long short-term memory model, which is a special type of artificial recurrent neural network architecture to predict Covid-19 cases after imposing lockdown during the second wave.

Chapter 6 presents an outline of epilepsy diagnosis by studying conventional neurodiagnostic techniques such as electroencephalography (EEG), magnetic resonance imaging, and other imaging modalities, along with different phases of seizures. Furthermore, this chapter reviews the automation of early diagnosis and seizure detection by integrating ML and DL techniques, which help in pattern recognition from EEG signals, reducing manual effort, improving diagnostic reliability, and facilitating personalized neurological treatment strategies.

Chapter 7 provides a cohesive structure is where ML and DL models are trained to learn the nonlinear relationships between geometrical parameters, switching states, and key performance metrics of resonant frequency, impedance bandwidth, gain, and radiation patterns to assist in fast synthesis and multi-objective optimization of reconfigurable patch structures using PIN diodes, varactors, radio frequency-microelectromechanical systems, and advanced materials.

 | https://doi.org/10.1515/9783112211991-202

Chapter 8 develops a mathematical model to deduct leakage present in a water pipeline network, as clean water is a main requirement for our society. The model is developed using augmented equation of continuity and the equation of head loss. The numerical values of the flow rate and pressure head are obtained using Python software for different values of Reynolds number values.

Chapter 9 focuses on the role of ML, DL, and hybrid models for accurately identifying financial distress in an organization. The study shows that by capturing complex nonlinear, temporal, and contextual patterns and integrating diverse data modalities, hybrid models overcome the limitations of traditional approaches.

Chapter 10 studies the transformative impact of AI-driven optimization techniques in domains such as industrial automation, energy systems, transportation, healthcare, and environmental management. The study examines how decision intelligence systems are used by AI and how they enhance the resource management, system efficiency and operation stability in different environments.

In Chapter 11, Dijkstra's graph theory approach is used to analyze Google Maps' route optimization. The study uses Python implementations and manual validation to show the reliability and effectiveness of Dijkstra's technique by modeling the road networks of Punjab and Uttarakhand as weighted graphs.

This book is essential for undergraduates, postgraduates, researchers, and professionals seeking to integrate AI into mathematical practices. It offers a foundational understanding for beginners, advanced techniques for experts, and insights into cutting-edge research, fostering interdisciplinary collaboration to solve complex real-world challenges.

Contents

List of Contributors

Hitesh Kumar
Department of Mathematics
Zakir Hussain College
University of Delhi
Delhi
India

Babita Bulwal
Department of Operational Research
University of Delhi
Delhi
India

Adarsh Anand
Department of Operational Research
University of Delhi
Delhi
India

Ompal Singh
Department of Operational Research
University of Delhi
Delhi
India

Rahul Sharma
Heidelberg University, Germany

Akanksha Gupta
Department of Computer Science
Graphic Era (Deemed to be University)
Dehradun
Uttarakhand
India

Kamesh Sonti
Sri Vasavi Engineering College
Pedatadepalli 534101
Andhra Pradesh, India

T. V. N. L. Aswini
Sri Vasavi Engineering College
Pedatadepalli 534101
Andhra Pradesh, India

Prathima M. G.
Department of CSE
Bangalore Institute of Technology
Bangalore
Karnataka
India

T. J. V. Subrahmanyeswara Rao
Sasi Institute of Technology and Engineering
Tadepalligudem 534101
Andhra Pradesh
India

Purnima K. Sharma
Chitkara University Institute of Engineering
and Technology
Chitkara University
Rajpura 140417
Punjab
India

Ajay Sudhir Bale
Department of ECE
New Horizon College of Engineering
Bengaluru 560103
Karnataka
India

Deepak N. R.
Department of Information Science
and Engineering
Atria Institute of Technology
Bengaluru
Karnataka
India

Chandramouli H.
Department of CSE
East Point College of Engineering
and Technology
Bengaluru
Karnataka
India

 | https://doi.org/10.1515/9783112211991-204

Naveen Ghorpade
Department of CSE
Vijay Vittala Institute of Technology
Bengaluru
Karnataka
India

Mamta B. Savadatti
Department of ECE
New Horizon College of Engineering
Bengaluru 560103
Karnataka
India

Varun Yogi Shantakumar
Department of Electrical and Electronic Engineering
California State University
Sacramento, CA
USA

Omkar K. Kanetkar
Department of ECE
New Horizon College of Engineering
Bengaluru 560103
Karnataka
India

Vasu Thakur
Department of Data and Computer Science
Ruprecht Karl University of Heidelberg
Heidelberg
Germany

Vikas Kumar Roy
Department of Data and Computer Science
Ruprecht Karl University of Heidelberg
Heidelberg
Germany

Ankush Mittal
Department of Data and Computer Science
College of Engineering Roorkee
Roorkee
Uttarakhand
India

Nupur Goyal
Department of Mathematics
Graphic Era (Deemed to be University)
Dehradun
Uttarakhand
India

Garvita Arora
Department of Computer Science
Graphic Era (Deemed to be University)
Dehradun
Uttarakhand
India

Akansha Gupta
Department of Computer Science
Graphic Era (Deemed to be University)
Dehradun
Uttarakhand
India

Dinesh Sharma
Department of ECE2
CCET (Degree Wing)
Panjab University,
Chandigarh 160019,
Chandigarh, India

Mullangi Vinod Kumar
Department of ECE
Sri Vasavi Engineering College
JNTUK
Andhra Pradesh, India

Navpreet
Department of ECE
Chitkara University Institute of Engineering and Technology
Chitkara University
Rajpura 140401
Punjab
India

Rakesh Chandra Bhadula
Department of Mathematics
Graphic Era Hill University
Dehradun
Uttarakhand
India

V. N. Kala
Department of Mathematics
GB Pant Engineering College
Pauri Garhwal
Uttarakhand
India

Jaipal Singh
Department of Mathematics
DBS (PG) College
Dehradun
Uttarakhand
India

Chandan Singh Ujarari
Department of Mathematics
Graphic Era Hill University
Dehradun
Uttarakhand
India

Dr Divya Sharma
Department of Computer Science Engineering
New Horizon College of Engineering
Bengaluru 560103
Karnataka
India

Dr Kushalatha M. R.
Department of ECE
Nitte Meenakshi Institute of Technology
Nitte (Deemed to be University)
Bengaluru 560064
Karnataka
India

Varun Yogi Shanthakumar
Department of Electrical and Electronic Engineering
California State University
Sacramento, CA
USA

Nikhil S. Nandi
Department of Mechanical Engineering
CMR University
Bengaluru
Karnataka
India

Astha Singh
Department of Mathematics
Graphic Era Hill University
Dehradun
Uttarakhand
India

Ashish Bisht
Department of Mathematics
Graphic Era Hill University
Dehradun
Uttarakhand
India

Deepanshu Tyagi
Department of Mathematics
Graphic Era Hill University
Dehradun
Uttarakhand
India

R. K. Tripathi
Department of Mathematics
Graphic Era Hill University
Dehradun
Uttarakhand
India

Hitesh Kumar, Babita Bulwal, Adarsh Anand, Ompal Singh

Chapter 1
A Three-Stage Numerical Analysis Approach for Predictive Modeling of Interconnected Online Engagement Metrics

Abstract: In various scientific fields, highly complex systems often consist of multiple interacting components, whose collective behavior is difficult to understand by analyzing individual components separately. Similar interaction-driven dynamics are also observed in modern digital platforms, where measurable system variables evolve in a coupled and nonlinear manner. In this study, YouTube is used as a representative complex system, and a dynamic multi-stage modeling framework is proposed to analyze the interactions among three interdependent system variables, namely views, likes, and subscribers for a given video. The proposed framework is inspired by epidemiology-based modeling concepts (SIR modeling), where information diffusion and user response are treated as a flow-driven process among different system states. The resulting mathematical formulation leads to a system of nonlinear differential equations, for which closed-form analytical solutions are not possible. Therefore, a fourth-order Runge-Kutta numerical scheme is employed to investigate the system behavior and the temporal evolution of the interacting variables. The study highlights the applicability of numerical and computational modeling techniques, commonly used in physical and engineering systems, for analyzing complex, interaction-driven systems beyond traditional content domains.

Keywords: Social media, YouTube, views, likes, subscribe, fourth-order RK method

1.1 Introduction

Social media has become an integral part of the modern social system, significantly influencing patterns of information dissemination, commercial discussions, and

---

Hitesh Kumar, Department of Mathematics, Zakir Hussain College, University of Delhi, India,
e-mail: hitesh.kmath@zh.du.ac.in
Babita Bulwal, Department of Operational Research, University of Delhi, Delhi, India,
e-mail: babitabulwal@gmail.com
Adarsh Anand, Department of Operational Research, University of Delhi, Delhi, India,
e-mail: adarsh.anand86@gmail.com
Ompal Singh, Department of Operational Research, University of Delhi, Delhi, India,
e-mail: drompalsingh1@gmail.com

 | https://doi.org/10.1515/9783112211991-001

everyday lifestyles. In recent years, an increasing number of individuals across all countries have shifted from traditional occupations to full-time content creation on social media platforms [1, 2]. This transition is driven by the emergence of social media as a sustainable, response-generating ecosystem, where creators can generate social, mental, and emotional value for society and earn income from their new creations.

Among various social media platforms, YouTube has its own distinct identity and has been effectively adopted within the social system [3]. YouTube is a platform on which content creators can earn income relatively easily. It allows creators to upload content independently while generating free visibility through the displayed growth of their videos [4]. Since the magnitude of displayed growth is sufficiently strong to attract views to a video, YouTube plays a significant role in content dissemination.

The behavior of a YouTube channel can be understood as a multi-stage process [5]. When a creator publishes content, the YouTube algorithm, depending on the type of video, initially shows the content to potential viewers within the social system through recommendations [6]. If these potential viewers watch the video and help spread it, the platform's recommendation algorithm further delivers it to other potential viewers, particularly promoting it among users interested in similar content. Among these nonsubscribers, only a few respond positively to the video and become viewers who like it, and from these users who like the video, an even smaller subset may eventually subscribe to the channel [7]. This observation indicates that the progression from viewing to liking and then subscribing gives rise to a structured, multi-stage process.

From this, it can be seen that for a YouTube channel, views, likes, and subscribers represent interdependent variables that jointly determine the success of the channel within the social system. A video must be viewed before it can spread, and a channel is generally subscribed to only after effective engagement with its content. Therefore, these variables cannot be treated independently; rather, they should be considered mutually dependent and analyzed within a multi-stage mathematical modeling framework. Motivated by this observation, the present study focuses on developing a mathematical model that captures the multi-stage interactions among views, likes, and subscribers in a dynamic system setting.

To ensure clarity and ease of understanding, the authors have structured the research article in a stepwise manner by dividing it into well-defined sections. The **Introduction** presents the importance of the study in a real-world context and clearly outlines the research objectives.

Section 1.2, which comprises the literature review, discusses several existing studies in the relevant research domain. Additionally, a comparative table is used to highlight the distinctions between the existing literature and the proposed work, thereby emphasizing the novelty and contribution of the present research.

Section 1.3 describes the basic assumptions and notations employed throughout the study. In **Section 1.4**, the proposed mathematical models are formulated and explained in a clear and systematic manner. Since, to the best of the authors' knowledge, some of the proposed models do not admit analytical solutions using traditional methods, **Section 1.5** elaborates on the numerical solution methodology adopted to obtain the model solutions.

Section 1.6 is devoted to data analysis. As the number of views, likes, and subscribers for a specific YouTube video is not publicly available and can only be accessed through the channel administrator, the authors have utilized real data on views, likes, and subscribers from a YouTube video for the purpose of this study.

Section 1.7, which focuses on validation, is divided into three subsections. The first subsection presents the model validation, which constitutes a crucial component of the study. In the second subsection, various statistical tools are employed to validate the proposed models. During the validation process, it was observed that the obtained results were insufficient to rank the models directly. Therefore, in **Section 1.7.3**, the authors applied the TOPSIS technique to rank the proposed models.

Finally, **Section 1.8** presents the conclusions of the study, while **Section 1.9** discusses the scope and potential directions for future research.

1.2 Literature Review

In recent years, extensive research has been carried out to understand the role and influence of social media on society. Early studies largely relied on qualitative methods to analyze potential viewer behavior and social impact. However, due to rapid technological developments, increased user interaction, and the availability of numerical data, mathematical and computational modeling approaches are increasingly being adopted to study social media dynamics [8].

One of the important early quantitative contributions was made by Mitchell and Kates (2009) [26], who employed Hawkes processes to model user activity related to online videos [9]. Their work laid the foundation for stochastic and event-based modeling in social media research. Subsequently, Rizoiu et al. [10] presented a significant and effective study on video popularity, identifying two key mechanisms through which content reaches potential audiences [10]. The first mechanism involves external publicity, where creators actively promote their content, while the second involves internal dissemination driven by audience interaction and word-of-mouth influence.

Despite these advances, much of the existing literature focuses on individual metrics such as views, likes, or subscribers. Limited attention has been paid to their joint and sequential interaction within an integrated framework. A summary highlighting the distinct features of the proposed modeling approach, along with a comprehensive review of the related literature, is provided in Table 1.1.

Table 1.1: Literature review and novelty of proposed work.

Research	Metrics studies	Methodology/model	Key relevance to proposed study
Szabo et al. [11]	Views (and votes as proxy for early engagement)	Early-time prediction of long-term popularity by statistical modeling (log-transform correlations)	Predict the long-term popularity of the online content.
Borghol et al. [12]	Popularity evaluation of YouTube videos	Time-series modeling for view rate dynamics; popularity curves	Modeling the popularity of user-generated YouTube videos, the proposed work also considers user and promotion-based popularity.
Richier et al. [13]	Epidemic-inspired mathematical modeling	In ODE-based mathematical modeling, an analytical solution exists.	Behaviors of viewing videos, which is connected to the proposed work due to ODE-based modeling.
Rizoiu et al. [14], (April)	Views including external promotions	Hawkes intensity process	Capture two-stage popularity (promotion-popularity. Closely related to the proposed multi-stage framework.
Hoiles et al. [15]	Views-subscribers	Sensitivity analysis for meta-level features that influence view count.	This chapter studies the interdependence between views and subscribers. The proposed work also considers the same.
Bärtl et al. [16]	Views, subscribers and channel growth	Large-scale statistical descriptive analysis	This chapter helps to understand the baseline of views and subscribers over time.
Babbar et al. [17]	View count and impact of word of mouth	ODE-based mathematical modeling	This chapter provides a strong understanding of subscribers with the effect of views.
Sargsyan et al. [18]	Half-life of the NEWS video on the YouTube channel	Statistical analysis and predictive modeling	This research study early-stage diffusion of a YouTube video, which helps to understand the diffusion process.
Cheng et al. [19]	Cascades/popularity (implicit views/ engagement)	Neural ODE + probabilistic diffusion	This research uses popularity prediction and uses the RK method.
Proposed work	Multi-stage diffusion dynamics considering views-likes-subscribers	ODE-based mathematical modeling with a lack of analytical solution	Combine study and prediction in all stages, using a real data set in all stages. Used the RK method in the absence of an analytical solution.

1.3 Assumptions and Notations

1.3.1 Assumptions

1. In this chapter, it is assumed that potential viewers first watch the video; after watching the video, they like it, and it is also assumed that only those viewers who like the video subscribe to the channel, and they do so directly through the video.
2. During this research, it is assumed that throughout the entire lifetime of the video, the number of potential viewers remain fixed.

1.3.2 Notations

m	=	Total number of potential viewers.
$N_v(t), N_l(t), N_s(t)$	=	Views, likes, and subscribers till time t, respectively.
$b_v(t), b_l(t), b_s(t)$	=	General representation for the rate of views, likes, and subscribers at time t, respectively.
b	=	Rate of views, likes, and subscribers at time t, respectively, in case of the same rates.
β	=	Learning parameter for views, likes, and subscribers at time t, respectively, in case of the same learning parameter.
b_v, b_l, b_s	=	Rate of views, likes, and subscribers at time t, respectively.
$\beta_v, \beta_l, \beta_s$	=	Learning parameter for views, likes, and subscribers at time t.
a_l	=	Proportion of the viewers who can be potential likers for the video.
a_l	=	Proportion of the likers who can be potential subscribers for the video.

1.4 Model Development

YouTube video views, likes, and subscribers are three different things that are interconnected with each other. So, model development has become a multi-stage process in which the first stage is for viewers, the second stage is for likes, and the third stage indicates the subscribers.

Stage 1: Viewer Dynamics

Suppose m is the fixed count of potential viewers for any video. Potential viewers can view a video via direct notifications from YouTube or recommendations from YouTube, or any other individual. Learning adoption rate considers both the direct factor (via notifications) from YouTube and the indirect factor (via sharing and recommendations) from YouTube. Therefore, the viewers' process can be studied by considering the learning adoption rate. Additionally, the rate of viewing a video depends on the remaining number of viewers. In this way, a mathematical equation to study YouTube views can be expressed as follows:

$$\frac{dN_v(t)}{dt} = b_v(t)(m - N_v(t)):N_v(0) = 0$$

where $b_v(t)$ is the rate of viewing the video by the remaining number of potential viewers.

Stage 2: Liking Dynamics

After watching the video, the viewer can either like or dislike the video, or they may simply switch to another video without giving any response. Therefore, it can be said that only a proportion a_l of viewers will take action to like the video. Hence, the rate at which the video gets likes depends on the remaining viewers who haven't liked the video yet but may do so in the future. Thus, the pattern of liking the video can be understood using the equation given below:

$$\frac{dN_l(t)}{dt} = b_l(t)(a_l N_v(t) - N_l(t)):N_l(0) = 0$$

where $b_l(t)$ is the rate of liking the video by the remaining number of potential likers.

Stage 3: Subscriber Dynamics

Not everyone who watches a video likes it, and not everyone who likes a video subscribes to the channel. So, only a certain part of the people who liked the video (let's call it a_s) might subscribe after liking it. That means the number of subscribers gained from a video depends on how many people are remaining to subscribe, who could potentially subscribe after liking the video. We can use the equation below to understand how subscribers increase from a video:

$$\frac{dN_s(t)}{dt} = b_s(t)(a_s N_l(t) - N_s(t)):N_s(0) = 0$$

where $b_s(t)$ indicates the rate of subscription to the YouTube channel after liking the video. Different mathematical models can be constructed by considering various combinations of rates. Here the authors consider two types of rates: the constant rate and the logistic growth rate. Both types of rates are very common and useful to study the behavior of viewers, likers, and subscribers. Combinations obtained by considering the rate functions are given in Table 1.2.

After solving the mathematical modeling framework using different combinations of rates, it can be observed that some have analytical solutions, while many

Table 1.2: Different rates that can be followed by the proposed modeling.

Model	$b_V(t)$	$b_l(t)$	$b_s(t)$
1	b	b	b
2	b	b	$\frac{b}{1+\beta e^{-bt}}$
3	b	$\frac{b}{1+\beta e^{-bt}}$	$\frac{b}{1+\beta e^{-bt}}$
4	$\frac{b}{1+\beta e^{-bt}}$	$\frac{b}{1+\beta e^{-bt}}$	$\frac{b}{1+\beta e^{-bt}}$
5	$\frac{b}{1+\beta e^{-bt}}$	b	b
6	$\frac{b}{1+\beta e^{-bt}}$	$\frac{b}{1+\beta e^{-bt}}$	b
7	b_V	b_l	b_s
8	b_V	b_l	$\frac{b_s}{1+\beta_s e^{-b_s t}}$
9	b_V	$\frac{b_l}{1+\beta_l e^{-b_l t}}$	$\frac{b_s}{1+\beta_s e^{-b_s t}}$
10	$\frac{v}{1+\beta_V e^{-b_V t}}$	$\frac{b_l}{1+\beta_l e^{-b_l t}}$	$\frac{b_s}{1+\beta_s e^{-b_s t}}$
11	$\frac{b_V}{1+\beta_V e^{-b_V t}}$	b_l	b_s
12	$\frac{b_V}{1+\beta_V e^{-b_V t}}$	$\frac{b_l}{1+\beta_l e^{-b_l t}}$	b_s

mathematical modeling frameworks do not contain analytical solutions using traditional methods. Table 1.3 has the solutions of different sets of models.

As observed from Table 1.3, analytical solutions can be derived for some of the proposed sets of mathematical models. However, it is also evident that for certain modeling frameworks, obtaining an analytical solution using traditional techniques is not feasible. In the domain of social media research, it has generally been assumed that analytical solutions must be obtainable for proposed mathematical models. However, this study demonstrates that for several of the methods proposed, analytical solutions are not achievable. Moreover, it is observed that those mathematical models for which analytical solutions are not obtainable tend to capture real-life scenarios more accurately. For instance, while Models 1 to 9 (not 3) yield analytical solutions, Model 9–12 (including 3) – which assume the rate of likes and subscribers follows a logistic growth pattern – do not offer an analytical solution. Yet, based on the

Table 1.3: Solution set of different modeling frameworks.

Model	$b_v(t)$	$b_l(t)$	$b_s(t)$
1	$m\left(1-e^{-bt}\right)$	$ma_l\left(1-(1+bt)e^{-bt}\right)$	$m\left(1-\left(1+bt+\frac{b^2t^2}{2}e^{-bt}\right)\right)$
2	$m\left(1-e^{-bt}\right)$	$ma_l\left(1-(1+bt)e^{-bt}\right)$	$\frac{m}{1+\beta e^{-bt}}\left(1-\left(1+bt+\frac{b^2t^2}{2}e^{-bt}\right)\right)$
3	$m\left(1-e^{-bt}\right)$	$\frac{ma_l}{1+\beta e^{-bt}}\left(1-(1+bt)e^{-bt}\right)$	No analytical solution
4	$m\left(\frac{1-e^{-bt}}{1+\beta e^{-bt}}\right)$	$\frac{a_l m}{1+\beta e^{-bt}}\left(1-e^{-bt}-(1+\beta)e^{-bt}\log\frac{e^{bt}+\beta}{\beta+1}\right)$	$\frac{a_l a_s m}{1+\beta e^{-bt}}\left(1-e^{-bt}-(1+\beta)e^{-bt}\log\frac{e^{bt}+\beta}{\beta+1}-\frac{\beta+1}{2}e^{-bt}\left(\log\frac{e^{bt}+\beta}{\beta+1}\right)^2\right)$
5	$m\left(\frac{1-e^{-bt}}{1+\beta e^{-bt}}\right)$	$ma_l\left[1-(1+(1+\beta)bt)e^{-bt}+(1+\beta)e^{-bt}Ln\left(\frac{1+\beta}{1+\beta e^{-bt}}\right)\right]$	$a_l a_s m\left(1-e^{-bt}-(1+\beta)e^{-bt}\log\frac{e^{bt}+\beta}{\beta+1}-\frac{\beta+1}{2}e^{-bt}\left(\log\frac{e^{bt}+\beta}{\beta+1}\right)^2\right)$
6	$m\left(\frac{1-e^{-bt}}{1+\beta e^{-bt}}\right)$	$\frac{a_l m}{1+\beta e^{-bt}}\left(1-e^{-bt}-(1+\beta)e^{-bt}\log\frac{e^{bt}+\beta}{\beta+1}\right)$	$a_l a_s m\left(1-e^{-bt}-(1+\beta)e^{-bt}\log\frac{e^{bt}+\beta}{\beta+1}-\frac{\beta+1}{2}e^{-bt}\left(\log\frac{e^{bt}+\beta}{\beta+1}\right)^2\right)$
7	$m\left(1-e^{-b_v t}\right)$	$ma_l\left(1-\frac{1}{b_l-b_v}\left(b_l e^{-b_v t}-b_v e^{-b_l t}\right)\right)$	$m\left(1-e^{-b_s t}+\frac{b_l b_s}{(b_s-b_v)(b_l-b_v)}\left(e^{-b_v t}-e^{-b_s t}\right)+\frac{b_v b_s}{(b_s-b_l)(b_l-b_v)}\left(e^{-b_l t}-e^{-b_s t}\right)\right)$
8	$m\left(1-e^{-b_v t}\right)$	$ma_l\left(1-\frac{1}{b_l-b_v}\left(b_l e^{-b_v t}-b_v e^{-b_l t}\right)\right.$	$\frac{m}{1+\beta_s e^{-b_s t}}1-e^{-b_s t}+\frac{b_l b_3}{(b_s-b_v)(b_l-b_v)}\left(e^{-b_v t}-e^{-b_3 t}\right)+\frac{b_v b_s}{(b_s-b_l)(b_l-b_v)}\left(e^{-b_l t}-e^{-b_s t}\right))$
9	$m\left(1-e^{-b_v t}\right)$	$\frac{ma_l}{1+\beta_2 e^{-b_2 t}}\left(1-e^{-b_2 t}-\frac{1}{b_2-b_v}\left(e^{-b_v t}-e^{-b_2 t}\right)\right)$	No analytical solution

10	$m\frac{1-e^{-b_1 t}}{1+\beta_1 e^{-b_1 t}}$	No analytical solution	No analytical solution
11	$m\frac{1-e^{-b_1 t}}{1+\beta_1 e^{-b_1 t}}$	No analytical solution	No analytical solution
12	$m\frac{1-e^{-b_1 t}}{1+\beta_1 e^{-b_1 t}}$	No analytical solution	No analytical solution

underlying equations and supporting literature, it is evident that the constant rate can be easily retrieved from the logistic rate by assuming $\beta=0$.

The absence of an analytical solution should not imply that the proposed mathematical model is ineffective or irrelevant. Therefore, the authors have employed an important approach – commonly referred to as *numerical methods* – to address those models for which analytical solutions are not available. Numerical methods are highly effective tools that allow us to obtain approximate numerical solutions without solving mathematical problems analytically. These methods can be used to solve various types of problems, such as root finding, interpolation, integration, and numerical differentiation.

1.5 Solution Methodology

As discussed earlier, due to the nonexistence of analytical solutions, approximate numerical solutions can be obtained using numerical methods. Picard's method, Euler's method, improved Euler's method, modified Euler's method, the Runge-Kutta method of second order (RK-2), and the Runge-Kutta method of fourth order (RK-4) are some of the commonly used techniques for obtaining numerical solutions [20]. Among all these methods, it is evident from the literature that the RK-4 method provides more accurate results compared to the others. Therefore, in this article, the RK-4 method has been used to obtain the numerical solution [21, 22].

In this chapter, the authors discuss three simultaneous differential equations. Let us consider the general form of the discussed differential equations:

$$\frac{dN_v(t)}{dt} = f(t, N_v(t), N_l(t), N_s(t)) : N_v(0) = 0$$

$$\frac{dN_l(t)}{dt} = g(t, N_v(t), N_l(t), N_s(t)) : N_l(0) = 0$$

$$\frac{dN_s(t)}{dt} = j(t, N_v(t), N_l(t), N_s(t)) : N_s(0) = 0$$

The numerical solution of the above simultaneous differential equations can be found using the following equations:

$$N_v(t_0 + h) = N_v(t_0) + \frac{1}{6}(v_1 + 2v_2 + 2v_3 + v_4)$$

$$N_l(t_0 + h) = N_l(t_0) + \frac{1}{6}(l_1 + 2l_2 + 2l_3 + l_4)$$

$$N_s(t_0 + h) = N_s(t_0) + \frac{1}{6}(s_1 + 2s_2 + 2s_3 + s_4)$$

where

$$
\begin{aligned}
v_1 &= hf(t_0, N_v(0), N_l(0), N_s(0) \\
v_2 &= hf\left(t_0 + \frac{h}{2}, N_v(0) + \frac{v_1}{2}, N_l(0) + \frac{l_1}{2}, N_s(0) + \frac{s_1}{2}\right) \\
v_3 &= hf\left(t_0 + \frac{h}{2}, N_v(0) + \frac{v_2}{2}, N_l(0) + \frac{l_2}{2}, N_s(0) + \frac{s_2}{2}\right) \\
v_4 &= hf(t_0 + h, N_v(0) + v_3, N_l(0) + l_3, N_s(0) + s_3) \\
l_1 &= hg(t_0, N_v(0), N_l(0), N_s(0) \\
l_2 &= hg\left(t_0 + \frac{h}{2}, N_v(0) + \frac{v_1}{2}, N_l(0) + \frac{l_1}{2}, N_s(0) + \frac{s_1}{2}\right) \\
l_3 &= hg\left(t_0 + \frac{h}{2}, N_v(0) + \frac{v_2}{2}, N_l(0) + \frac{l_2}{2}, N_s(0) + \frac{s_2}{2}\right) \\
l_4 &= hg(t_0 + h, N_v(0) + v_3, N_l(0) + l_3, N_s(0) + s_3) \\
s_1 &= hj(t_0, N_v(0), N_l(0), N_s(0) \\
s_2 &= hj\left(t_0 + \frac{h}{2}, N_v(0) + \frac{v_1}{2}, N_l(0) + \frac{l_1}{2}, N_s(0) + \frac{s_1}{2}\right) \\
s_3 &= hj\left(t_0 + \frac{h}{2}, N_v(0) + \frac{v_2}{2}, N_l(0) + \frac{l_2}{2}, N_s(0) + \frac{s_2}{2}\right) \\
s_4 &= hj(t_0 + h, N_v(0) + v_3, N_l(0) + l_3, N_s(0) + s_3)
\end{aligned}
$$

Now, the authors have a way to find numerical solutions in place of analytical ones. However, another problem arises when considering parameter estimation. Most software tools, such as SPSS, LINGO, Maple, Mathematica, and MATLAB, typically require analytical solutions for estimating parameters. To address this challenge, Microsoft Excel offers a simple and effective alternative. The authors estimated the parameters by minimizing the mean absolute error (MAE) across all datasets. Specifically, they first calculated the MAE between actual and predicted values for views, likes, and subscribers. In the next step, they summed all three MAEs to obtain a total error value. Finally, using Excel's Solver tool, they minimized this total error by adjusting the values of the model parameters [23, 24].

1.6 Data Analysis

Whenever a video is uploaded, it starts receiving views from its potential viewers and is liked by some viewers, and some viewers who like the video subscribe to it. Data related to video views and the count of likes can be easily collected from the YouTube video by anyone. But the number of subscribers gained by the video is not visible to everyone. The data related to views, likes, and subscribers can be collected from the

YouTube channel itself, which is available only to the owners of the YouTube channel. That's why it has become difficult to collect data from all three stages.

The dataset belongs to an entertainment video. The dataset can be directly compiled in an excel sheet, although the viewing pattern is illustrated in Figure 1.1. During data collection, it was evident in YouTube data that most viewers (approximately 97%) accessed the video through YouTube search results. This indicates that views generated through sharing were very low. The behavior of the viewers suggests that the likes and subscribers gained from this video were primarily due to YouTube search, implying that the video had a significant impact on its audience. The views were recorded on a day-by-day basis.

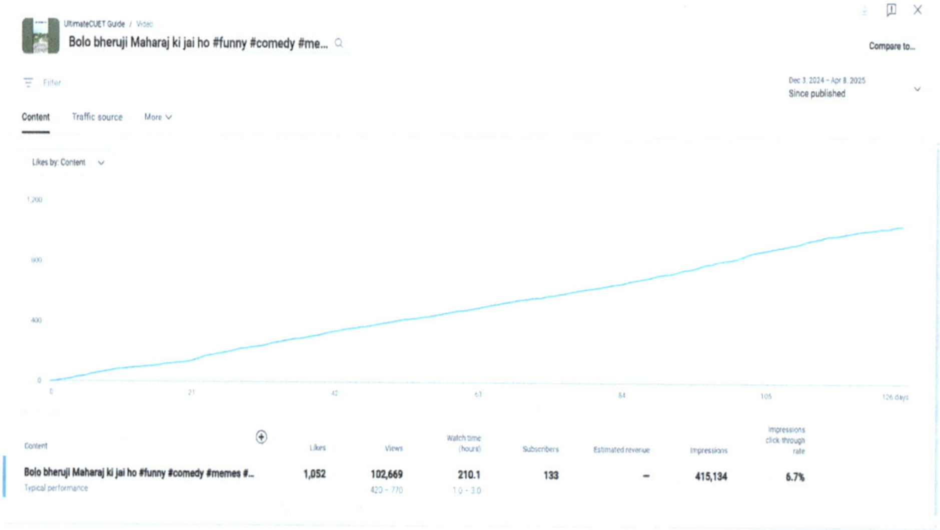

Figure 1.1: View data of YouTube video (DSI).

After watching the video, potential viewers decide whether to like, dislike, or ignore it. In the mathematical modeling framework, the authors focus only on the viewers who liked the video. Therefore, data on video likes was collected along the same timeline as shown in Figure 1.2. Here as well, the number of likes was recorded on a day-by-day basis and is presented as numerical counts.

As discussed in the modeling work, it can be assumed that viewers who like the video are likely to subscribe to the YouTube channel. In other words, if an individual subscribes to the channel, there is a high probability (95–100%) that they first liked the video. Therefore, the authors collected subscriber data along the same timeline as shown in Figure 1.3.

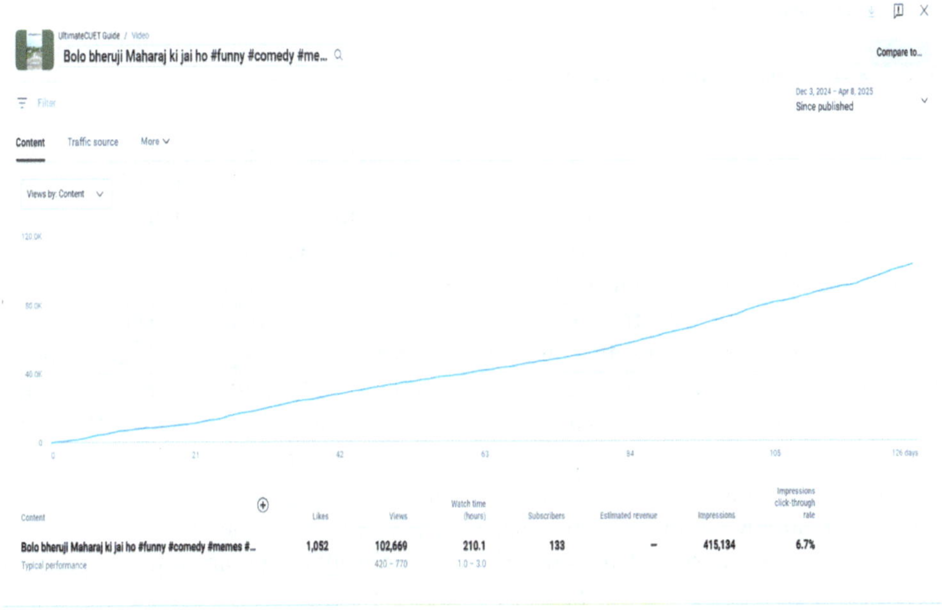

Figure 1.2: Likes data of YouTube video (DSII).

Figure 1.3: Subscribers data of YouTube video (DSIII).

1.7 Validation

1.7.1 Parameter Estimation

To validate the proposed modeling framework on the above-discussed dataset, the first problem occurs due to the nonexistence of an analytical solution. Therefore, as discussed above, softwares like SPSS, MATLAB, Mathematica, and LINGO, are not useful. To the best of the authors' knowledge, the only viable method utilizes Excel, where numerical solutions can be applied and values of different parameters can be estimated easily. Here authors discuss the method of parameter estimation using Excel.

Parameter estimation using tools such as SPSS and LINGO is generally convenient. However, in scenarios where an analytical solution is not obtainable, these software packages may fail to provide accurate parameter estimation. In such cases, Microsoft Excel can serve as an effective alternative for parameter estimation.

When using Excel, the process begins with generating an approximate numerical solution by employing suitable numerical methods. This step is carried out while carefully considering the role of parameters in the underlying mathematical model. All parameters used in the numerical solution must be assigned to specific fixed cells in the Excel sheet. These designated cells are then used to input initial parameter values, based on available data and the mathematical modeling framework.

Subsequently, the error between the predicted and actual values is computed using various statistical tools. In this study, the authors have employed the MAE as the error metric. Once the MAE is calculated, Excel's Solver tool is utilized for optimization.

The optimization process involves selecting an objective function – in this case, the minimization of the MAE. Since the objective function depends on the parameter values, the next step is to specify the cells representing these parameters in the "by changing variable cells" section of the Solver. If the parameters are subject to any constraints, these constraints can also be incorporated at this stage.

Finally, the nonlinear least squares method is selected within Solver to minimize the objective function. Upon execution, Solver provides the set of parameter values for which the MAE is minimized. Estimated values of all parameters are given in Table 1.4.

1.7.2 Model Validation

Since the proposed mathematical model is a three-stage framework, it is important to note that it is quite rare to find datasets that capture the process occurring at each stage of a multi-stage model. However, in this study, data corresponding to all three stages – views, likes, and subscribers – are available. Therefore, the proposed mathematical model has been analyzed and compared across all three stages. The authors use MAE, relative absolute error (RAE), mean bias deviation (MBD), mean absolute percentage error (MAPE), and R^2 for comparison of the models.

Table 1.4: Parameter values for all data models.

Model	m	b_v	b_l	b_s	β_v	β_l	β_s	a_l	a_s
1	6546649	0.00011	0.00011	0.00011	–	–	–	0.02899	0.39205
2	198536.7	0.004115	0.004115	0.004115	–	–	8.7461	0.065818	0.85665
3	59837.64	0.059586	0.059586	0.059586	–	8.288054	8.288054	0.015348	0.15366
4	58106.5	00063091	0.063091	0.063091	8.005876	8.005876	8.005876	0.015531	0.153432
5	10292245	0.0065224	0.0065224	0.0065224	125.452	–	–	0.0468288	0.6155554
6	318191	0.00229	0.00229	0.00229	0.00352	0.00352	–	0.456	0.361
7	1307810	0.000542	0.315359	0.488203	–	–	–	0.012236	0.13003
8	1292912	0.00055	0.307912	0.515564	–	–	1.9721	0.0122255	0.129537
9	93393643	0.000074	0.38640	0.89	–	0.19736	0.156956	0.0121074	0.1278815
10	2925886	0.003642	0.95	0.985	16.2334	0.2	0.355356	0.011605	0.126308
11	1025049049	0.00001644	0.269	0.356	1.450038	0.00534	–	0.0128728	0.126137
12	19074248	0.00287	0.7580389	0.5024313	86.83369	–	–	0.011782	0.127967

Table 1.5: Model validation on stage 1 (views of YouTube video).

Model	MAE	RAE	MBD	MAPE	R^2
1	3495.919	0.146338	599.2733	10.81588	0.986655
2	5502.097	0.230315	705.7964	18.08975	0.969361
3	9168.782	0.383801	4004.652	17.66411	0.82832
4	10062.06	0.421193	3887.187	19.96061	0.801022
5	**1098.937**	**0.046001**	170.6558	4.267863	**0.998086**
6	4758.524	0.19919	2658.689	12.53414	0.978221
7	3724.241	0.155895	815.8876	11.36972	0.985145
8	3727.721	0.156041	660.3539	11.573	0.985115
9	3479.189	0.145637	598.0425	10.75735	0.986762
10	1836.613	0.07688	**21.4925**	**5.64666**	0.994934
11	3449.304	0.144386	1095.441	10.03145	0.987021
12	1980.284	0.082894	234.4182	5.9201	0.99435

On observing Table 1.5, where authors try to rank the performance, it can be observed that directly ranking is not possible because different criteria provide different rankings.

Table 1.6: Model number performance decreasing order.

Criteria	Model number performance decreasing order											
MAE	5	10	12	11	9	1	7	8	6	2	3	4
RAE	5	10	12	11	9	1	7	8	6	3	2	4
MBD	10	5	12	9	1	8	2	7	11	6	4	3
MAPE	5	10	11	12	9	1	7	8	6	3	2	4
R^2	5	10	12	11	9	1	7	8	6	2	3	4

Based on Table 1.6, it can be said that direct comparison and performance measures are not possible. To overcome this issue authors used the MCDM technique to provide comparisons. Technique for order preference by similarity to ideal solution (TOPSIS) is a widely used method in multi-criteria decision-making. When each model has different types of performance values, using TOPSIS becomes very convenient. In cases where a model performs well on one metric but poorly on another, this method helps to create a balance between them. One of the best features of TOPSIS is that it does not require assigning any weights manually. It works entirely based on data and operates with complete objectivity. Therefore, in this chapter, TOPSIS has been used to evaluate model performance. The results obtained through TOPSIS will be discussed after analyzing the performance for subscribers.

If we look at the various performance criteria obtained from the likes data, we find that different models perform best under different criteria. The rankings of these models can also be seen in Table 1.7.

Table 1.7: Model validation on stage 2 (likes of YouTube video).

Model	MAE	MAE	MBD	MAPE	R^2
1	497.465	1.921961	493.1825	98.3759	0.9555
2	91.61626	0.35396	61.27414	34.37516	0.972373
3	80.63223	0.311523	47.18116	34.09332	0.955405
4	80.70834	0.311817	46.82225	32.82938	0.946133
5	118.8618	0.459224	76.39689	40.23806	0.950625
6	1196.896	4.624219	−1168.12	172.0214	0.965162
7	14.99889	0.057948	3.104947	6.200805	0.996309
8	15.08847	0.058294	**1.134316**	6.299961	0.996308
9	13.8436	0.053485	1.378656	5.675827	0.996896
10	**11.88324**	0.045911	3.549813	**5.183085**	**0.998696**
11	18.85787	0.072858	−14.4222	7.476567	0.997117
12	11.40463	**0.044062**	3.451066	5.272526	0.99873

Table 1.8: Ranking of matrix based on likes.

Criteria	Model Number performance decreasing order											
MAE	5	10	12	11	9	1	7	8	6	2	3	4
RAE	5	10	12	11	9	1	7	8	6	3	2	4
MBD	10	5	12	9	1	8	2	7	11	6	4	3
MAPE	5	10	11	12	9	1	7	8	6	3	2	4
R^2	5	10	12	11	9	1	7	8	6	2	3	4

As mentioned in Table 1.8, the ranking of the models is not clearly defined. Therefore, in such a situation, we will use TOPSIS for comparison. The results obtained through TOPSIS will be discussed after analyzing the performance for subscribers as given in Table 1.9.

Table 1.9: Model Validation on stage 3 (subscribers of YouTube Video).

Model	MAE	RAE	MBD	MAPE	R^2
1	63.2888	1.939858	−62.743	99.1925171	0.857982
2	22.53048	0.69058	−17.5658	54.5189155	0.883437
3	13.37092	0.409831	−12.0049	43.2901483	0.952281
4	14.08362	0.431675	−12.5439	45.037337	0.94428
5	23.96433	0.734528	−18.224	56.4599175	0.86669
6	24.0592	0.737436	−21.4116	57.1627653	0.871951
7	1.647788	0.050506	−0.44968	8.68396307	0.997641
8	1.6562	0.050764	−0.33923	8.8802902	0.997599
9	**1.537149**	**0.047115**	**−0.32967**	**7.68086814**	0.997858
10	1.9621	0.060324	−1.31411	7.69125	0.9978
11	1.96809	0.058248	−0.99589	7.99016652	0.99783
12	2.109656	0.064663	−1.52989	9.26985759	**0.997965**

If we carefully examine the results obtained from the subscribers' dataset, it becomes evident that the modeling performance cannot be directly ranked. As shown in Table 1.10, the rankings of the models vary under different performance criteria.

Table 1.10: Ranking of matrix based on likes.

Criteria	Model Number performance decreasing order											
MAE	9	7	8	10	11	12	3	4	2	5	6	1
RAE	9	7	8	10	11	12	3	4	2	5	6	1
MBD	9	8	7	11	10	12	3	4	2	5	6	1
MAPE	9	10	11	7	8	12	3	4	2	5	6	1
R^2	12	9	10	11	7	8	3	4	2	5	6	1

Since a straightforward ranking is not feasible, the authors have used the TOPSIS method for ranking. In the table given below, all competing models are ranked according to their TOPSIS scores.

Table 1.11: Ranking of models on different datasets.

Dataset -model	1	2	3	4	5	6	7	8	9	10	11	12
DSI	12	9	7	8	10	11	3	2	1	5	4	6
DSII	11	9	7	8	10	12	5	4	1	3	6	2
DSIII	6	10	11	12	2	9	7	8	5	1	3	4

Upon closely analyzing the results obtained through an MCDM technique, TOPSIS, the authors observe that if a model secures Rank 1 on Dataset I (DSI), its ranking tends to drop on Dataset II (DSII) or Dataset III (DSIII), and a completely different model starts performing at Rank 1. Therefore, a direct comparison of model performance is not possible [25]. Performing at rank 1 as given in Table 1.11.

To find the best performance among the proposed modeling frameworks, the authors use a direct but effective method by taking the average of model performances as given in Table 1.12.

Table 1.12: Final modeling ranking.

Model	1	2	3	4	5	6	7	8	9	10	11	12
Average of ranking	9.66	9.33	8.33	9.33	7.33	10.66	5	4.667	2.33	3	4.33	4
Ranking	11	9 = 10	8	9 = 10	7	12	6	5	1	2	4	3

From the above ranking, it can be observed that Model 9, Model 10, Model 12, Model 11, and Model 8 have the first five ranks, and the mathematical modeling framework proposed in these models lacks analytical solutions in the second and third stages. (Note: only model 8 has an analytical solution in the second stage, but it takes the 5th position.) In this way, it can be said that in the absence of an analytical solution, a numerical method can be applied to study real-life situations.

1.8 Conclusion

Based on the mathematical modeling framework presented above, it can be concluded that the number of subscribers, likes, and views on YouTube videos can be easily understood using the multi-stage modeling approach proposed in this chapter. It can also be observed that if the analytical solution of a mathematical model cannot be obtained using traditional methods, it does not imply the failure of the model itself. In the absence of an analytical solution, numerical methods can be effectively employed. As established in several studies, the Runge-Kutta method of fourth order yields better results compared to other numerical techniques. Therefore, it can be stated that the fourth-order RK method is one of the most accurate and low-error approaches for solving such models.

Furthermore, based on the results obtained from the comparison of methods, it can also be concluded that the rates of views, likes, and subscribers – and the impact of word-of-mouth – vary in the context of YouTube videos. These variations could only be effectively captured and demonstrated through the use of numerical methods.

1.9 Future Implements

The mathematical modeling presented in this chapter – despite the absence of an analytical solution – demonstrates that such models can still be effectively utilized in the domain of social media and OTT platforms. This approach may catalyze future advancements, significantly encouraging the development and application of many more mathematical models in this field.

References

[1] Saini, R., & Chourasiya, A. (2023). The shifting landscape of social media and its implications for the career path of content creators. *International Journal of Advanced Research in Commerce, Management & Social Science (IJARCMSS)*, *6*(01), 01–10.

[2] Vasilieva, E. N., Vasiliev, E. S., & Poltavskaya, M. B. (2020, March). Transformation of economic activity in the digital era: Practices for generating income by young people in social networks. In *Institute of Scientific Communications Conference* (pp. 691–700). Cham: Springer International Publishing.

[3] Burgess, J., & Green, J. (2018). *YouTube: Online Video and Participatory Culture*. John Wiley & Sons.

[4] Hou, M. (2019). Social media celebrity and the institutionalization of YouTube. *Convergence, 25*(3), 534–553.

[5] Anand, A., Babbar, G., Aggrawal, N., & Ram, M. (2024). Modelling dynamic nature of YouTube's digital ecosystem and its analysis using numerical methods. *International Journal of System Assurance Engineering and Management, 15*, 1–11.

[6] Zhou, R., Khemmarat, S., & Gao, L. (2010, November). The impact of YouTube recommendation system on video views. In *Proceedings of the 10th ACM SIGCOMM Conference on Internet Measurement* (pp. 404–410).

[7] Dobrian, F., Awan, A., Joseph, D., Ganjam, A., Zhan, J., Sekar, V., . . . & Zhang, H. (2013). Understanding the impact of video quality on user engagement. *Communications of the ACM, 56*(3), 91–99.

[8] Rajput, J. K., Patel, K., Srivastava, M., & Kumar, V. (2024). Special issue on exploring material science and technology innovations. *Prabha Materials Science Letters, 4*(1), 1–3.

[9] Mitchell, L., & Cates, M. E. (2009). Hawkes process as a model of social interactions: A view on video dynamics. *arXiv Preprint arXiv:0907.3864.*

[10] Rizoiu, M.-A., Xie, L., Sanner, S., Cebrian, M., Yu, H., & Van Hentenryck, P. (2016). Expecting to be HIP: hawkes intensity processes for social media popularity. *arXiv Preprint.*

[11] Szabo, G., & Huberman, B. A. (2010). Predicting the popularity of online content. *Communications of the ACM, 53*(8), 80–88.

[12] Borghol, Y., Mitra, S., Ardon, S., Carlsson, N., Eager, D., & Mahanti, A. (2011). Characterizing and modelling popularity of user-generated videos. *Performance Evaluation, 68*(11), 1037–1055.

[13] Richier, C., Altman, E., Elazouzi, R., Altman, T., Linares, G., & Portilla, Y. (2014). Modelling view-count dynamics in youtube. *arXiv Preprint arXiv:1404.2570.*

[14] Rizoiu, M. A., Xie, L., Sanner, S., Cebrian, M., Yu, H., & Van Hentenryck, P. (2017, April). Expecting to be hip: Hawkes intensity processes for social media popularity. In *Proceedings of the 26th International Conference on World Wide Web* (pp. 735–744).

[15] Hoiles, W., Aprem, A., & Krishnamurthy, V. (2016). Engagement dynamics and sensitivity analysis of YouTube videos. *arXiv Preprint arXiv:1611.00687.*

[16] Bärtl, M. (2018). YouTube channels, uploads and views: A statistical analysis of the past 10 years. *Convergence, 24*(1), 16–32.

[17] Babbar, G., Anand, A., & Aggrawal, N. (2024). Modelling & analyzing view growth pattern of YouTube videos inculcating the impact of subscribers, word of mouth and recommendation systems. *International Journal of Mathematical, Engineering & Management Sciences, 9*(3) 435–450.

[18] Sargsyan, A., Dutta, H. S., & Pfeffer, J. (2025). Half-life of Youtube news videos: Diffusion dynamics and predictive factors. *arXiv Preprint arXiv:2507.21187.*

[19] Cheng, Z., Zhou, F., Xu, X., Zhang, K., Trajcevski, G., Zhong, T., & Yu, P. S. (2024). Information cascade popularity prediction via probabilistic diffusion. *IEEE Transactions on Knowledge and Data Engineering, 36*, 8541–8555.

[20] Aggrawal, D., Anand, A., Kumar, H., Singht, O., & Singh, I. (2024). Multi-stage multi-generation based product adoption: An analysis using numerical methods. *Nonlinear Studies, 31*(2), 1–17.

[21] Delic, G. (2025). Numerical solution of ordinary differential equations. In *Guide to Numerical Algorithm Design and Development: Including Legacy Examples from Fortran and MathCAD in High Precision (pp. 197–232).* Cham: Springer Nature Switzerland.

[22] Kumar, H., Singh, O., Anand, A., & Irshad, M. S. (2023). Studying multi-stage diffusion dynamics using epidemic modeling framework. *International Journal of Mathematical, Engineering and Management Sciences*, *8*(1), 105.

[23] Barati, R. (2013). Application of excel solver for parameter estimation of the nonlinear Muskingum models. *KSCE Journal of Civil Engineering*, *17*(5), 1139–1148.

[24] Singh, O., Kumar, H., Anand, A., & Kapur, P. K. (2025). Modelling technology diffusion process under dynamic market scenario. In *Data-centric Approaches to Industrial Decisions: Technology, Digitisation and Business Decisions* (pp. 357–390). Singapore: Springer Nature Singapore.

[25] Singh, D. P., Saini, S., Avikal, S., & Prasad, B. (2024). An mcdm-based approach for the selection of natural fiber for marine applications. *Prabha Materials Science Letters*, *4*(1), 96–106.

[26] Mitchell, L., & Cates, M. E. (2010). Hawkes process as a model of social interactions: a view on video dynamics. Journal of Physics A: Mathematical and Theoretical, *43*(4), 045101.

Rahul Sharma, Akanksha Gupta

Chapter 2
Predictive Modeling with AI: Methods, NLP Applications, and Deployment

Abstract: Predictive modeling utilizes historical data and machine learning algorithms to forecast future outcomes, serving as a cornerstone of modern data-driven decision-making. This chapter provides an expert-level examination of the end-to-end predictive modeling life cycle, focusing heavily on Natural Language Processing (NLP) as a crucial modality for extracting actionable insights from unstructured text. We delineate the practical boundaries between traditional machine learning and deep learning, trace the architectural evolution of sequence models from recurrent neural networks to modern self-attention transformers, and explore rigorous data pipeline engineering principles to combat critical pitfalls such as data leakage and training-serving skew. Furthermore, we examine empirical evaluation strategies suited for imbalanced data—advocating for precision-recall curves over standard ROC curves, probability calibration techniques, and cost-sensitive learning functions. Finally, we address operational requirements through the lens of MLOps, analyzing distribution shift, algorithmic fairness, and governance frameworks, reinforced by real-world case studies in healthcare, finance, and customer service.

Keywords: Predictive Modeling, Natural Language Processing, Transformer Models, MLOps, Data Leakage, Model Calibration, Cost-Sensitive Learning, Algorithmic Fairness, Feature Stores

2.1 Introduction

Predictive modeling, the practice of using historical data and machine learning (ML) algorithms to forecast future outcomes, has become a cornerstone of modern data-driven decision-making [1]. A model learns patterns from past data and applies this knowledge to new inputs to classify events, estimate trends, or forecast future values. This chapter provides a comprehensive, expert-level overview of the predictive modeling life cycle, with a specialized focus on natural language processing (NLP) as a critical modality for extracting insights from unstructured text.

Rahul Sharma, Department of Mathematics and Computer Science, Heidelberg University, e-mail: rahul.sharma@stud.uniheidelberg.de
Akanksha Gupta, Department of Computer Science, Graphic Era (Deemed to be University), Dehradun, Uttarakhand, India, e-mail: akanshagupta.cse@geu.ac.in

 | https://doi.org/10.1515/9783112211991-002

This chapter begins by establishing the foundational concepts of ML, distinguishing between supervised and unsupervised learning paradigms and clarifying the practical differences between traditional ML and deep learning (DL). It then charts the complete, iterative life cycle of a predictive model, from initial problem definition to post-deployment monitoring. A significant portion is dedicated to NLP, exploring the essential preprocessing steps of tokenization and embedding, and tracing the architectural evolution from recurrent neural networks (RNNs) to the revolutionary transformer models that power today's large language models (LLMs).

Crucially, this chapter argues that successful predictive modeling is an engineering discipline, not just an algorithmic one. It emphasizes the necessity of robust data and feature pipelines, highlighting common but critical pitfalls such as data leakage and training-serving skew. A pragmatic tour of model families – from interpretable linear models and high-performance tree ensembles to the DL arsenal of CNNs and transformers – is presented through the lens of practical trade-offs between accuracy, speed, and interpretability.

Furthermore, this chapter advocates for empirical rigor in model evaluation. It moves beyond simple accuracy to discuss the selection of appropriate metrics for imbalanced classification problems, such as the preference for precision-recall curves over receiver operating characteristic (ROC) curves. It also delves into the importance of model calibration to ensure that a model's confidence scores are trustworthy and the alignment of evaluation with business value through cost-sensitive learning.

Finally, the discussion extends to the nonfunctional requirements of modern artificial intelligence (AI) systems. It addresses the challenges of maintaining model robustness in the face of distribution shifts, ensuring algorithmic fairness by detecting and mitigating bias, and managing risk through comprehensive AI governance. The principles of machine learning operations (MLOps) are detailed as the framework for operationalizing these practices, covering continuous integration and continuous delivery (CI/CD), automated monitoring, and safe rollback strategies. Real-world case studies in healthcare, finance, and customer service ground these concepts in practical application, illustrating both successes and failure modes. This chapter concludes with a look toward the future, exploring emerging trends like multimodal models, retrieval-augmented generation (RAG), and the drive for smaller, more efficient models, and provides a detailed checklist for practitioners to guide the development and deployment of predictive models responsibly and effectively.

2.2 Foundations of Predictive Modeling

Predictive modeling is the process of analyzing historical data to train models that can predict future or unknown outcomes [1]. It forms the core of predictive analytics, enabling data-driven forecasting and decision automation across a vast array of fields, from finance and marketing to healthcare and logistics. At its heart, a predictive model is a mathematical function that learns a mapping from a set of input features to a target outcome. The quality and utility of this function depend on the data it learns from, the algorithm used to learn, and the rigor of the process used to develop and validate it.

2.2.1 Defining the Predictive Task: Supervised, Unsupervised, and Time-Series Learning

ML models, the engines of predictive modeling, are typically categorized by the nature of the learning task they are designed to solve. These paradigms are not rigid silos but rather points on a spectrum, often combined in practice to build sophisticated systems.

Supervised Learning: This is the most common paradigm in predictive modeling. Models are trained on a dataset containing labeled examples, where both the input features (X) and the corresponding known target outputs (Y) are provided. The model's objective is to learn the mapping function f such that $Y \approx f(X)$. Supervised learning encompasses two primary task types:

- **Classification**: The target variable Y is a discrete category. The model predicts the class label for a given input. Examples include predicting whether an email is spam or not (binary classification) or classifying a customer support ticket into one of several categories (multi-class classification) [2].
- **Regression**: The target variable Y is a continuous numerical value. The model predicts a quantity. Examples include forecasting next quarter's sales revenue or predicting the price of a house based on its features [2].

Unsupervised Learning: In this paradigm, models are given data without explicit labels (X only) and must discover inherent patterns or structures within it. While not always "predictive" in the forecasting sense, unsupervised techniques are crucial for exploratory data analysis (EDA) and feature engineering. Common tasks include:

- **Clustering**: Grouping similar data points together based on their features. For instance, a k-means algorithm can segment customers into distinct behavioral groups without any predefined labels [2].
- **Dimensionality Reduction**: Compressing data into a lower-dimensional representation while preserving as much meaningful information as possible. Principal component analysis is a classic example.

Time-Series Forecasting: This is a specialized form of supervised learning where the data points are ordered chronologically. The temporal dependency between observations is a critical feature. Models designed for time-series data must account for patterns like trends, seasonality, and cyclicality. Traditional statistical methods include

Autoregressive Integrated Moving Average (ARIMA) and exponential smoothing, while modern approaches leverage ML algorithms, including sequence-based neural networks, which can learn complex temporal dynamics from historical data to predict future values [2].

In practice, these paradigms often intersect. For example, an unsupervised clustering algorithm might first be used to create a "customer segment" feature from unlabeled behavioral data. This new feature can then be fed into a supervised regression model to predict the lifetime value for each customer, demonstrating a powerful, hybrid approach that leverages the strengths of both paradigms.

2.2.2 Machine Learning Versus Deep Learning: A Practical Distinction

The terms ML and DL are often used interchangeably, but they represent a meaningful distinction in approach and capability.

ML is the broad, umbrella field encompassing any algorithm that enables a computer to learn from data without being explicitly programmed [3]. This includes a wide range of algorithms, from simple linear regression to complex ensemble models like gradient boosting. A defining characteristic of many "traditional" ML workflows is the reliance on *manual feature engineering*, where domain experts carefully craft informative input features from raw data to guide the model's learning process [2].

DL is a specific subfield of ML that utilizes multilayered artificial neural networks (often called "deep" neural networks) [4]. The key innovation of DL is its ability to perform *representation learning*: the model automatically learns a hierarchy of features directly from raw data. For instance, in image recognition, the initial layers of a deep network might learn to detect simple edges, subsequent layers might combine these to recognize textures and shapes, and the final layers might identify complex objects [2]. This capability to learn features end to end has led to breakthrough performance on unstructured data types like images, audio, and text.

This distinction introduces a fundamental tension that shapes much of modern AI development: the trade-off between performance and interpretability. Traditional ML models, such as linear regression or decision trees, are often more transparent. One can inspect the model's coefficients or rules to understand *why* a particular prediction was made [2]. DL models, with their millions of interconnected parameters, are frequently described as "black boxes" because their internal decision-making logic is opaque and difficult for humans to interpret directly [5]. This lack of transparency is not merely a technical curiosity; it is a primary driver of risk in high-stakes applications like healthcare and finance. The deployment of powerful but opaque DL models creates significant business, ethical, and regulatory challenges, which in turn has spurred the development of entire sub-fields dedicated to explainable AI, fairness auditing, and robust AI governance frameworks.

2.2.3 The End-to-End Model Life Cycle

Developing a predictive model is not a single event but an iterative, cyclical process that spans from conceptualization to long-term maintenance. Understanding this life cycle is essential for building models that are not only accurate but also reliable and sustainable in a production environment (see Figure 2.1).

The key stages of this life cycle are as follows:

1. **Problem Definition**: Clearly articulate the business objective and the prediction target. Define the success criteria, which may include a specific performance metric (e.g., accuracy) and a broader business key performance indicator (KPI) (e.g., reduction in customer churn).
2. **Data Collection and Understanding**: Gather relevant historical data from various sources. This stage involves EDA to understand the data's attributes, quality, and limitations.
3. **Data Preprocessing**: Clean the raw data by handling missing values, correcting inconsistencies, and removing outliers. The data is then split into training, validation, and test sets.
4. **Feature Engineering**: Transform raw data into informative features that the model can effectively learn from. This may involve creating new variables, encoding categorical data, or scaling numerical features.
5. **Model Selection and Training**: Choose a set of candidate algorithms appropriate for the task and data. The models are then trained on the training dataset, where they learn to adjust their internal parameters to map input features to the target outcome.
6. **Validation and Hyperparameter Tuning**: Evaluate the trained models on the validation set to assess their ability to generalize to unseen data. Hyperparameters – the model's configuration settings that are not learned from the data – are adjusted to optimize performance. This is an iterative loop with the training stage.
7. **Evaluation**: Once a final model is selected and tuned, its performance is measured on a completely independent hold-out test set. This provides an unbiased estimate of how the model will perform in the real world.
8. **Deployment**: The validated model is integrated into a production environment where it can receive new data and generate predictions to inform business decisions.
9. **Monitoring and Maintenance**: Continuously monitor the model's performance and the distribution of incoming data in the production environment. Over time, performance may degrade due to changes in the underlying data patterns, a phenomenon known as "drift."
10. **Feedback Loop**: Insights from monitoring feed back into the life cycle. If performance degrades, the model may need to be retrained on fresh data, or the entire problem may need to be redefined, thus restarting the cycle.

Figure 2.1: The iterative life cycle of a predictive model. This diagram illustrates the end-to-end process, emphasizing the continuous feedback loop from production monitoring back to the initial development stages. Source: Authors' own work.

2.3 NLP as a Predictive Modality

NLP is a subfield of AI that gives computers the ability to understand, interpret, and generate human language. In the context of predictive modeling, NLP unlocks the vast amount of information contained in unstructured text data – from customer reviews and social media posts to medical records and financial reports – and transforms it into a powerful predictive signal.

2.3.1 From Text to Tensors: Tokenization and Embeddings

Before a ML model can process text, the raw string of characters must be converted into a numerical format. This transformation involves two fundamental steps: tokenization and embedding.

Tokenization: This is the initial process of breaking down a continuous stream of text into smaller, discrete units called tokens [6]. These tokens can be words, characters, or, most commonly in modern NLP, subwords. For example, the sentence "AI drives innovation" might be tokenized into the list ["AI", "drives", "innovation"]. Advanced techniques like Byte-Pair Encoding (BPE) or WordPiece use subword tokenization, which breaks rare or complex words into more common components (e.g., "innovation" might become ["innov", "##ation"]). This approach effectively handles out-of-vocabulary words and creates a more manageable vocabulary size [7].

Embeddings: Once tokenized, each token is mapped to a high-dimensional numerical vector known as an embedding. An embedding is a learned representation that captures the token's semantic meaning [6]. In a well-trained embedding space, tokens with similar meanings will have vectors that are close to each other. For example, the vectors for "king" and "queen" will be closer than the vectors for "king" and "car." This allows the model to reason about semantic relationships mathematically, moving beyond simple one-hot encodings where every word is treated as an isolated, independent entity [6].

The choice of a tokenizer and its corresponding embedding space is not a trivial preprocessing step; it is a foundational architectural decision. The vocabulary and subword splitting rules of a chosen tokenizer, often inherited from a large pretrained model, define the model's fundamental understanding of language. Using a tokenizer trained on general web text for a highly specialized domain like legal or medical analysis can lead to suboptimal performance, as critical domain-specific terms may be fragmented into nonsensical subwords, effectively placing a ceiling on the model's potential before training even begins.

2.3.2 The Evolution of Sequence Models: From RNNs to Transformers

Language is inherently sequential; the order of words matters. Consequently, NLP has long relied on models designed to process sequential data.

Recurrent Neural Networks (RNNs): Architectures like long short-term memory and gated recurrent units were the dominant approach for sequence modeling for many years. RNNs process a sequence token by token, maintaining an internal hidden state that acts as a memory, carrying context from previous tokens to inform the processing of the current one [3]. However, RNNs face two major limitations: they struggle to capture long-range dependencies in very long sequences due to the "vanishing gradient problem," and their sequential nature makes them difficult to parallelize during training, slowing down development [3].

The Transformer Architecture: Introduced by Vaswani et al. [8] in the seminal paper "Attention Is All You Need," the transformer architecture revolutionized NLP [9]. It completely dispensed with recurrence and instead introduced the *self-attention mechanism*. Self-attention allows the model to weigh the relevance of every other token in the sequence when processing a given token, all at once [10]. This parallel processing capability, combined with its superior ability to model long-range dependencies, made the transformer vastly more efficient to train on large datasets and more powerful than its recurrent predecessors.

The self-attention mechanism can be understood as a form of automated, dynamic feature engineering that occurs within the model. For each token at each layer, the attention mechanism constructs a new, context-rich representation by computing a weighted average of all other tokens in the sequence. This is fundamentally different from an RNN, which builds its context sequentially. A transformer re-engineers the contextual representation for every token at every layer, allowing it to capture highly complex and nuanced relationships. This powerful mechanism is the core engine behind the recent explosion in LLMs like the GPT series and Bidirectional Encoder Representations from Transformers (BERT), which have achieved state-of-the-art performance across a wide range of NLP tasks [11].

2.3.3 Framing NLP Tasks as Predictive Problems

Many practical NLP applications can be framed as standard predictive modeling tasks, enabling the use of the life cycle and evaluation techniques discussed previously.

- **Text Classification**: Predicting a single categorical label for an entire piece of text. This is one of the most common NLP tasks in industry. Examples include:
  - *Sentiment Analysis*: Predicting whether a product review is positive, negative, or neutral

- *Spam Detection*: Predicting if an email is spam or not
- *Topic Categorization*: Assigning a news article to a category like "Sports," "Politics," or "Technology"

- **Sequence Labeling**: Predicting a label for each individual token in an input sequence. A key application is *named entity recognition*, which involves identifying and classifying entities like person names, organizations, and locations within a text [7].
- **Sequence-to-Sequence (Seq2Seq) Tasks**: Predicting an entirely new output sequence based on an input sequence. These tasks often use an encoder-decoder architecture, where an encoder processes the input sequence into a contextual representation, and a decoder generates the output sequence from that representation. Examples include:
  - *Machine Translation*: Translating a sentence from one language to another
  - *Text Summarization*: Generating a concise summary of a long document
  - *Question Answering*: Providing a textual answer to a question based on a given context paragraph
- **Language Modeling**: The fundamental task of predicting the next token in a sequence given the preceding tokens. This is the core task for generative models like GPT and is used for applications like autocomplete, code generation, and creative writing [12].

The modern paradigm for tackling these tasks is *transfer learning*, which involves a two-stage process. First, a large model like BERT or GPT is pretrained on a massive, unlabeled text corpus (e.g., the entire internet) using a self-supervised objective like language modeling [7]. This pretraining phase endows the model with a deep, general understanding of language, grammar, and world knowledge. Second, this pretrained model is *fine-tuned* on a much smaller, task-specific labeled dataset. This approach has fundamentally altered the economics of NLP development. The immense cost of pretraining is borne by a few large organizations, while a wide array of practitioners can achieve state-of-the-art results with modest data and computational resources by fine-tuning existing models. This democratization of advanced NLP has shifted the primary challenge for many from model building to the responsible application, evaluation, and operationalization of these powerful tools.

2.4 Data and Feature Pipelines for Robust Models

No matter how sophisticated the algorithm, a predictive model's performance is fundamentally constrained by the quality of the data it is trained on. The maxim "garbage in, garbage out" holds especially true in ML [13]. A well-architected data and feature pipeline is therefore not an auxiliary component but the foundational infrastructure upon which reliable and effective models are built. This pipeline encompasses the entire journey of data from its raw state to the clean, informative features that a model consumes.

2.4.1 The Critical Path: Data Collection, Cleaning, and Feature Engineering

The initial stages of the model life cycle are dedicated to preparing the data, a process that is often the most time-consuming yet most impactful part of a project.

- **Data Collection and Integration**: The process begins by gathering data from all relevant sources, which may include structured databases, event logs, user-generated content, or third-party Application Programming Interfaces (APIs). It is crucial to ensure that the collected data is representative of the scenarios the model will encounter in production. Data from different sources often needs to be joined and synchronized, requiring careful handling of timestamps and identifiers to maintain data integrity [69].
- **Data Cleaning**: Real-world data is invariably messy. This stage involves a systematic process of identifying and rectifying issues such as missing values (which may be imputed or rows/columns removed), inconsistent entries (e.g., standardizing categorical labels like "USA" and "United States"), duplicates, and outliers. Data cleaning procedures must be designed carefully to avoid introducing biases or leaking information [13].
- **Feature Engineering**: This is the art and science of transforming raw data into features that better represent the underlying problem for the model. For structured, tabular data, this remains a critical, often manual, process driven by domain expertise. It can include creating new features (e.g., deriving "day of the week" from a timestamp), encoding categorical variables into a numerical format (e.g., one-hot encoding), and scaling numerical features to a common range. For unstructured data like text or images, DL models have automated much of this process, but initial data representation still matters greatly [69].

2.4.2 A Silent Killer: Identifying and Preventing Data Leakage

Data leakage is one of the most insidious and common failure modes in predictive modeling. It occurs when information that would not be available at the time of prediction is inadvertently used to train the model [14]. This leads to a model that appears to have outstanding performance during validation but fails catastrophically when deployed in the real world because it has learned to "cheat" by using information it will not have access to.

There are two primary forms of data leakage:

- **Target Leakage**: This happens when a feature in the training data is a direct or indirect result of the target variable itself. A classic example is attempting to predict whether a patient has pneumonia and including a feature took_antibiotic_medicine. Since antibiotics are typically prescribed *after* a pneumonia diagnosis, this feature leaks the outcome to the model, leading to artificially perfect predictions that are useless in a real diagnostic setting [15].
- **Train-Test Contamination**: This subtler form of leakage occurs when information from the validation or test set bleeds into the training process. A common mistake is to perform data preprocessing steps, such as fitting a scaler or an imputer for missing values, on the entire dataset *before* splitting it into training and test sets. This allows the training process to learn from the statistical properties (e.g., mean and variance) of the test set, violating its status as truly "unseen" data [14].

Data leakage is best understood not as an algorithmic error but as a failure of the operational development process. It reveals a breakdown in the crucial task of simulating the production prediction environment during training. The remedies are therefore procedural, not algorithmic. The most critical principle is to *always perform the train-test split before any data-dependent preprocessing*. For time-series data, this means using strict chronological splits, training on the past to predict the future [15]. Every feature should be scrutinized with the question: "Would I have this exact information at the moment I need to make a prediction?" If the answer is no, the feature must be excluded or re-engineered.

2.4.3 Consistency and Scale: The Role of Feature Stores

As ML systems scale within an organization, a significant operational challenge emerges: ensuring that the features used for model training are computed in the exact same way as the features used for real-time inference. Discrepancies between these two pipelines, known as *training-serving skew*, are a primary cause of production model failure [68].

A *feature store* is an architectural pattern designed specifically to solve this problem [16]. It is a centralized data management layer for ML features, typically comprising two main components [17]:

- **Offline Store**: A repository for historical feature data, often built on a data warehouse or data lake. It is designed for high-throughput access to large volumes of data and is used to generate training datasets.
- **Online Store**: A low-latency database (e.g., a key-value store like Redis) that serves the most recent feature values for real-time inference.

By providing a single source of truth for feature definitions and computation logic, a feature store acts as a formal contract between the data engineering teams that produce features and the ML teams that consume them. When a feature (e.g., user_average_purchase_value_last_30_days) is defined, the logic is implemented once. The feature store's infrastructure then ensures this logic is used to populate both the offline store for training and the online store for serving, thereby eliminating training-serving skew by design [67, 70]. This makes the feature store a critical architectural linchpin connecting the data engineering pipeline to the MLOps deployment pipeline, enabling reliable, scalable, and reusable feature management.

2.5 A Tour of Model Families and Their Trade-Offs

The landscape of predictive algorithms is vast, with no single model being universally superior – a concept formalized in the "No Free Lunch" theorem [18]. The art of model selection is therefore not a search for a silver bullet but a process of constraint optimization. The "best" model for a given problem is the one that optimally balances the specific constraints of the task, including the structure of the data, the required predictive performance, the budget for latency and computational cost, and the need for interpretability (summarized in Table 2.1).

2.5.1 Interpretable Baselines: Linear Models and Decision Trees

Before turning to more complex models, it is a best practice to establish a strong baseline with simpler, more interpretable algorithms:

- **Linear Models**: This family includes linear regression for continuous targets and logistic regression for classification. They operate on the assumption that the output is a linear combination of the input features. Despite their simplicity, they are computationally efficient, fast to train, and highly interpretable. The influence of each feature on the prediction is directly captured by its learned coefficient, making it easy to explain the model's logic [3]. While they cannot capture complex, nonlinear relationships on their own, they can be surprisingly effective when combined with careful feature engineering.
- **Decision Trees**: These nonparametric models learn a hierarchy of if-else rules that partition the data to make a prediction. A single decision tree is highly interpretable, as one can trace the path of decisions from the root to a leaf node to understand a specific prediction. However, a single, deep tree is highly prone to overfitting the training data, capturing noise rather than the underlying signal [3].

2.5.2 The Power of Ensembles: Random Forests and Gradient Boosting

To overcome the limitations of single models, especially the high variance of decision trees, *ensemble methods* combine the predictions of multiple models to produce a more robust and accurate final prediction:

- **Random Forests**: This is a classic bagging (Bootstrap Aggregating) ensemble. It trains a large number of decision trees independently on different random subsets (bootstrap samples) of the training data and features. The final prediction is made by averaging the predictions (for regression) or taking a majority vote (for classification) of all the individual trees. This process effectively reduces variance and makes the model much less prone to overfitting than a single tree [19].
- **Gradient Boosting Machines (GBMs)**: This is a boosting ensemble method. Instead of training models independently, GBMs build them sequentially. Each new tree is trained to correct the errors (specifically, the residual errors) of the ensemble of trees that came before it. This iterative, error-correcting process allows GBMs to achieve extremely high predictive performance. Modern implementations like *XGBoost*, *LightGBM*, and *CatBoost* have become the state of the art and are often the winning choice for competitions and applications involving structured or tabular data [20].

2.5.3 The Deep Learning Arsenal: CNNs, Transformers, and Emerging Architectures

For unstructured data, DL models have become the undisputed state-of-the-art due to their ability to learn feature representations directly from the data:

- **Convolutional Neural Networks (CNNs)**: These are specialized for processing grid-like data, most notably images. They use layers of convolutional filters that slide across the input to detect local patterns, such as edges and textures in an image's early layers, and more complex shapes and objects in deeper layers. This exploitation of spatial locality makes them incredibly efficient and powerful for computer vision tasks [21].
- **Transformers:** As discussed in Section 2.3, transformers are the dominant architecture for sequential data, particularly text. Their self-attention mechanism allows them to model complex, long-range dependencies, making them ideal for a wide range of NLP tasks [8]. Their application is also expanding to other domains, including computer vision and time-series analysis.
- **Structured State-Space Models (SSMs)**: A newer class of deep sequence models, such as S4 and Mamba, is emerging as a potential successor or complement to the transformer. SSMs use principles from classical state-space models to process sequences with high computational efficiency, particularly for very long sequences

where the quadratic complexity of the transformer's self-attention can become a bottleneck [22].

Table 2.1: A practical comparison of model family trade-offs.

Model family	Best for	Performance	Interpretability	Data needs	Speed (inference)
Linear models	Linearly separable data, baselines	Moderate	High	Low	Very fast
Tree ensembles (GBM)	Structured/tabular data	High	Medium	Medium	Fast
CNNs	Images, grid-like data	Very high	Low	High	Medium-fast
Transformers	Sequences (text, time-series)	State-of-the-art	Very low	Very high	Slow

Source: Authors' own work.

2.6 Rigorous Evaluation and Validation Strategies

Building a predictive model is an empirical science. A model's true value is not determined by its performance on the data it was trained on, but by its ability to generalize and make accurate predictions on new, unseen data. Therefore, a rigorous and well-designed evaluation strategy is not an afterthought but a core component of the model development life cycle. This involves choosing the right validation scheme, selecting metrics that align with business goals, and ensuring the model's outputs are trustworthy.

2.6.1 Beyond Accuracy: Selecting the Right Metrics for the Job (ROC vs. PR Curves)

The most fundamental principle of model evaluation is the separation of data. A dataset is typically split into three parts:

- **Training Set**: Used to train the model's parameters
- **Validation Set**: Used to tune the model's hyperparameters and select the best-performing model architecture
- **Test Set**: A final, hold-out set that is used only once to provide an unbiased estimate of the final model's performance on unseen data

For classification tasks, simple *accuracy* (the percentage of correct predictions) can be highly misleading, especially in the presence of *class imbalance*. For example, in a fraud detection dataset where only 1% of transactions are fraudulent, a naive model that predicts "not fraud" for every transaction will achieve 99% accuracy, yet it is completely useless. More nuanced metrics are required:

- **Precision**: Of all the instances the model predicted as positive, what fraction were actually positive? (TP/(TP + FP)). High precision is critical when the cost of a false positive is high (e.g., flagging a legitimate transaction as fraud).
- **Recall (Sensitivity)**: Of all the actual positive instances, what fraction did the model correctly identify? (TP/(TP + FN)). High recall is critical when the cost of a false negative is high (e.g., failing to detect a cancerous tumor).
- **F_1-Score**: The harmonic mean of precision and recall, providing a single score that balances both metrics.

For models that output a probability score, their performance can be evaluated across all possible decision thresholds using visualization curves:

- **ROC (Receiver Operating Characteristic) Curve**: This curve plots the true positive rate (recall) against the false positive rate at various threshold settings. The *area under the curve (ROC-AUC)* is a single metric summarizing this trade-off. An AUC of 1.0 represents a perfect classifier, while an AUC of 0.5 represents a model with no discriminative ability [23].
- **Precision-Recall (PR) Curve**: This curve plots precision against recall at various thresholds. The *area under the PR curve (PR-AUC)* is the corresponding summary metric.

In scenarios with significant class imbalance, the *PR curve is often more informative than the ROC curve* [24]. The ROC curve's reliance on the false positive rate can present an overly optimistic view of performance, as a large number of true negatives can make the FPR appear low even with a substantial number of false positives. The PR curve, by focusing only on precision and recall, provides a more direct and sensitive assessment of the model's performance on the rare, positive class [25]. The choice of which metric to optimize is not merely a statistical one; it is an implicit statement of business priority. Opting for a metric like PR-AUC or F_1-score signals that the business objective is heavily weighted toward correctly identifying the minority class, accepting a potential trade-off in the form of more false alarms.

2.6.2 Trusting the Probabilities: The Importance of Model Calibration

For many applications, a model's class prediction is not sufficient; the associated probability or confidence score is also critical for decision-making. A model is considered *well-calibrated* if its predicted probabilities accurately reflect the true likelihood of the event [26]. For instance, if a model assigns a probability of 80% to 100 different predictions, we expect that approximately 80 of those predictions will be correct.

A significant issue with many modern, high-capacity models, particularly deep neural networks, is that while they may be highly accurate, they are often poorly calibrated, tending to be *overconfident* in their predictions [27]. An uncalibrated model that confidently predicts a 99% probability for an incorrect outcome can be dangerous and misleading in a production system.

Calibration can be assessed visually using *reliability diagrams* (calibration plots) and quantitatively with metrics like the *Brier score* or *expected calibration error* [28]. If a model is found to be miscalibrated, its outputs can be corrected post hoc using techniques such as:

- **Platt Scaling**: A simple method that trains a logistic regression model on the classifier's scores to map them to calibrated probabilities [29].
- **Isotonic Regression**: A more powerful, nonparametric method that fits a non-decreasing function to the scores [30].
- **Temperature Scaling**: A simple and effective technique for multi-class neural networks that rescales the model's pre-softmax outputs (logits) by a single learned "temperature" parameter to improve calibration without affecting accuracy [27].

Model calibration is the essential bridge between a model's raw output score and a probability that can be used for rational, risk-based decision-making. If a business rule states "intervene only if the predicted risk is above 75%," that threshold is only meaningful if the 75% probability is a reliable reflection of the true event frequency.

2.6.3 Aligning with Business Value: Cost-Sensitive and Utility-Based Evaluation

In the real world, not all prediction errors are created equal. A false negative in a medical diagnosis (missing a disease) can have far more severe consequences than a false positive (triggering an unnecessary follow-up test). Standard evaluation metrics often treat all errors with equal weight, which may not align with the actual business or clinical impact.

Cost-sensitive learning addresses this by explicitly incorporating the asymmetric costs of different errors into the model's training and evaluation process [31]. This is often achieved by defining a *cost matrix* that assigns a specific cost or utility to each cell of the confusion matrix (true positive, false positive, true negative, and false negative) [32]. The model's objective then becomes to minimize the total expected cost, rather than simply minimizing the number of errors.

This approach formalizes the business problem directly within the model's objective function. By modifying the loss function to penalize more costly errors more heavily – for example, by assigning a higher weight to the minority class in an imbalanced problem – the practitioner directly causes the model to learn a decision boundary that is optimized for business utility, not just statistical accuracy [33]. This represents a mature stage of predictive modeling, where the model transitions from being a passive predictor to an active, optimized decision-making agent.

2.7 Managing Robustness, Fairness, and Risk in AI Systems

As predictive models are increasingly deployed in critical, real-world systems, their evaluation must extend beyond raw predictive performance. A responsible AI practitioner must also ensure that models are robust to changing environments, are fair and equitable in their outcomes, and are deployed within a governance framework that manages privacy and security risks.

2.7.1 When the World Changes: Handling Distribution Shift and Concept Drift

The independent and identically distributed assumption – that training and test data are drawn from the same underlying distribution – is a cornerstone of classical ML theory [34]. However, in nearly every practical application, this assumption is violated [35]. The real world is nonstationary; user behavior evolves, economic conditions change, and the deployment of a model can itself alter the environment it operates in. This mismatch between the training data distribution and the production data distribution is known as *distribution shift*, and it is a primary cause of performance degradation in deployed models [13].

Key types of distribution shift include [36]:

- **Covariate Shift**: The distribution of the input features, $P(X)$, changes, but the relationship between features and the target, $P(Y|X)$, remains stable. For example, a loan default model trained on data from one demographic may perform poorly when applied to a new demographic with different income and credit score distributions, even if the underlying drivers of default are the same.
- **Concept Drift**: The fundamental relationship between the features and the target, $P(Y|X)$, changes over time. For example, a product recommendation model may become obsolete as customer preferences and trends evolve, making the features that once predicted a purchase no longer relevant.

Distribution shift is not an esoteric edge case; it is the default, inevitable state for any deployed model. This reality means that a "deploy and forget" strategy is a recipe for failure. The entire engineering discipline of MLOps is, in large part, a framework for combating the inevitability of distribution shift. Strategies for mitigation include [37]:

- **Monitoring**: Continuously tracking the statistical properties of incoming production data and comparing them to the training data to detect drift. Statistical tests like the Kolmogorov-Smirnov test or metrics like the population stability index are often used.
- **Retraining**: Periodically retraining the model on fresh, recent data to ensure it adapts to the latest patterns.
- **Domain Adaptation**: Employing techniques that explicitly try to adapt a model trained on a source distribution to perform well on a different but related target distribution.

2.7.2 Algorithmic Fairness: Detecting and Mitigating Unwanted Bias

Predictive models learn from historical data, and if that data reflects existing societal biases, the model will learn and potentially amplify those biases [38]. A hiring model trained on decades of biased hiring data might learn to unfairly penalize female candidates, or a loan approval model might learn to discriminate based on race by using proxy variables like zip code.

Addressing this requires a systematic approach to *algorithmic fairness*. This involves measuring, and then mitigating, unwanted bias in a model's outcomes across different demographic groups defined by sensitive attributes (e.g., race, gender, and age).

There is no single, universal definition of fairness; instead, there are multiple mathematical formalizations, which are often mutually exclusive (see Table 2.2) [39]. The choice of which fairness metric to optimize is a socio-technical decision that depends on the specific context and ethical goals.

Bias mitigation techniques can be applied at three stages of the modeling pipeline [38]:

- **Preprocessing**: Modifying the training data to remove bias, for example, by resampling or reweighing data to ensure balanced representation
- **In-processing**: Modifying the learning algorithm itself, for example, by adding a fairness constraint or a penalty term to the loss function that discourages disparate outcomes [40]
- **Postprocessing**: Adjusting the model's predictions after they have been made, for example, by setting different classification thresholds for different demographic groups to achieve a desired fairness metric.

Ultimately, fairness is a socio-technical problem, not a purely mathematical one. It requires a multidisciplinary approach involving data scientists, domain experts, ethi-

Table 2.2: A comparison of common group fairness definitions.

Metrics	Definition	Interpretation
Demographic parity	The model's predictions are independent of the sensitive attribute. $P(\hat{Y}=1\|A=0)=P(\hat{Y}=1\|A=1)$	The proportion of individuals receiving a positive outcome is the same across all groups.
Equal opportunity	The model has an equal true positive rate across groups. $P(\hat{Y}=1\|Y=1,A=0)=P(\hat{Y}=1\|Y=1,A=1)$	Among qualified individuals, the chance of receiving a positive outcome is the same across all groups.
Equalized odds	The model has equal true positive rates and equal false positive rates across groups.	Both qualified and unqualified individuals are treated equitably across groups, balancing the concerns of both equal opportunity and avoiding disparate harm.

Source: Authors' own work

cists, and affected communities to define what "fair" means in a given context. The role of the practitioner is to then implement, measure, and transparently report on the chosen fairness definition using tools like *Model Cards*, which document a model's performance characteristics, including its evaluation on different demographic subgroups [41].

2.7.3 Privacy, Security, and AI Governance Frameworks

The use of sensitive personal data for training predictive models raises significant privacy and security concerns. Models can sometimes memorize specific details from their training data, making them vulnerable to attacks that could expose private information [42].

Privacy-preserving machine learning is a field dedicated to developing techniques that allow for model training without compromising the privacy of the underlying data [43]. Key approaches include:

- **Differential Privacy (DP)**: A rigorous mathematical framework that provides formal privacy guarantees. It involves adding carefully calibrated statistical noise to the data or the learning algorithm, such that the output of the model is statistically indistinguishable whether or not any single individual's data was included in the training set [44].
- **Federated Learning (FL)**: A decentralized training approach where the data remains on local devices (e.g., mobile phones or hospital servers). A global model is trained by aggregating model updates (e.g., gradients) from these devices, without the need to centralize the raw, sensitive data [45].

These techniques introduce a fundamental trilemma between *privacy*, *utility* (model accuracy), and *cost* (computational complexity). Stronger privacy guarantees, such as those provided by DP, often come at the cost of reduced model accuracy due to the added noise. Similarly, techniques like FL and homomorphic encryption (computing on encrypted data) introduce significant engineering and computational overhead [45]. Navigating this trade-off is a key strategic challenge.

To manage these complex risks, organizations are increasingly adopting formal *AI governance frameworks*. This involves establishing processes like independent model validation, creating AI ethics review boards, maintaining thorough documentation, and ensuring mechanisms for user recourse and contestability of model decisions, as mandated by regulations like the EU's General Data Protection Regulation (GDPR) [46].

2.8 MLOps: Deploying and Maintaining Models in Production

Developing a high-performing predictive model is only the first step. Operationalizing that model – deploying it into a production environment where it can deliver value reliably, scalably, and sustainably – is a distinct and complex engineering challenge. *MLOps* is the discipline that applies Development and Operations (DevOps) principles to the ML life cycle to address this challenge [47]. MLOps is fundamentally a risk management framework, where practices like automation, monitoring, and versioning are implemented to mitigate the unique and continuous risks associated with production AI systems [13].

2.8.1 CI/CD for Machine Learning: Automating the Path to Production

At the core of MLOps is the automation of the end-to-end model life cycle through a CI/CD pipeline:

- **Version Control**: In an MLOps context, version control extends beyond just the source code. It is critical to version all components of the ML system: the *data* used for training, the model training *code*, and the resulting trained *model artifacts*. This comprehensive versioning is the bedrock of reproducibility, allowing any model to be recreated and audited, and enabling safe rollbacks to previous versions [48].
- **Continuous Integration (CI)**: The CI stage for ML involves automatically testing and validating any new changes. This goes beyond typical software unit tests to include data validation (checking for schema changes or drift), model validation

(ensuring the new model's performance on a test set meets a predefined threshold), and even fairness and bias checks [69].

- **Continuous Delivery (CD)**: Once a model has passed all CI checks, the CD stage automates its deployment into a production or staging environment. This ensures a rapid, reliable, and repeatable path to production, reducing manual errors and accelerating the delivery of improvements [69].
- **Continuous Training (CT)**: Many MLOps pipelines also incorporate CT, where the model retraining process is automatically triggered by events such as the availability of new data or the detection of performance degradation [69].

Figure 2.2 illustrates a typical architecture for a real-time model serving system, a common deployment target for an MLOps pipeline.

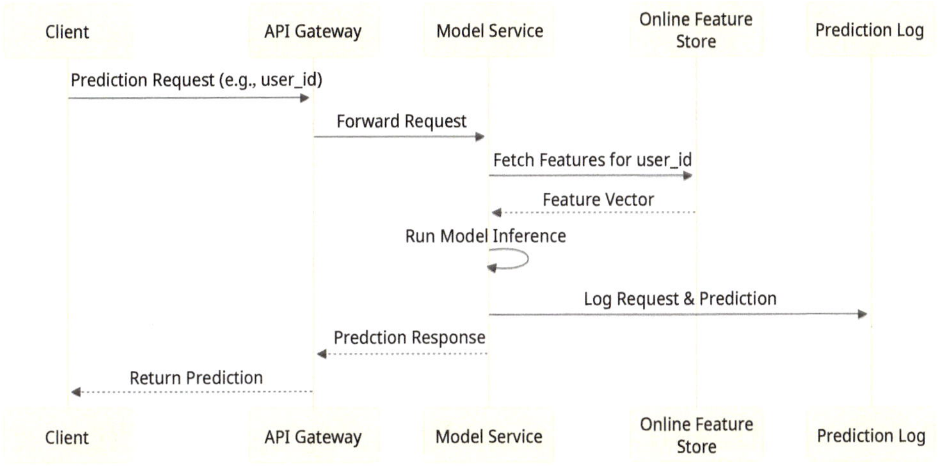

Figure 2.2: A real-time inference architecture. This diagram shows the flow of a prediction request from a client through an API to a model service, which retrieves real-time features from an online feature store before making a prediction. Source: Authors' own work.

2.8.2 Closing the Loop: Monitoring, Alerting, and Rollback Strategies

A deployed model is not a static artifact; it is a dynamic system that requires continuous oversight.

- **Monitoring**: Comprehensive monitoring is essential for maintaining the health and performance of a production model. Monitoring should cover multiple aspects [13]:
  - *Operational Health*: Tracking system metrics like server latency, throughput, and error rates

  - *Data Drift*: Monitoring the statistical distribution of incoming prediction data to detect shifts from the training data distribution
  - *Model Performance*: Tracking predictive performance (e.g., accuracy, AUC) over time, which requires a feedback loop to collect ground-truth outcomes
  - *Fairness*: Monitoring for any divergence in performance or outcomes across different demographic subgroups
- **Safe Deployment Strategies**: To mitigate the risk of deploying a new model that performs worse than its predecessor, safe rollout strategies are used:
  - *Shadow Deployment*: The new model is deployed alongside the existing one. It receives a copy of the live production traffic, and its predictions are logged for analysis but are not served to users. This allows for a direct comparison of the new and old models on live data without any user impact [69].
  - *Canary Deployment/A/B Testing*: The new model is initially rolled out to a small percentage of users. Its performance is closely monitored, and if it proves to be superior to the old model, traffic is gradually shifted until the new model is serving 100% of requests [69].
- **Rollback and Governance**: If monitoring detects a critical issue – such as a sudden drop in performance, a spike in errors, or the discovery of a harmful bias – a robust MLOps system must have a *rollback plan*. This allows the system to quickly and automatically revert to a previously known stable version of the model, minimizing the impact of the failure [69]. This entire process forms a continuous governance loop, ensuring that the AI system remains effective, reliable, and aligned with its objectives over its entire life cycle (see Figure 2.3).

2.9 Predictive AI in Action: Industry Case Studies

To ground the preceding theoretical concepts in practical reality, this section examines three real-world applications of predictive modeling, with a focus on NLP. These case studies highlight the methods used, the challenges encountered, and the tangible impact of deploying AI in complex business environments.

2.9.1 Healthcare: AI-Assisted Triage in Emergency Departments

Problem: In a busy hospital emergency department (ED), accurately and quickly triaging patients to determine their required level of care (e.g., discharge, general ward admission, or intensive care unit (ICU)) is a critical and high-pressure task. Mis-triage can lead to poor patient outcomes and inefficient resource allocation.

Solution: Researchers have developed predictive models that assist clinicians by analyzing both structured data (e.g., vital signs and age) and unstructured text from

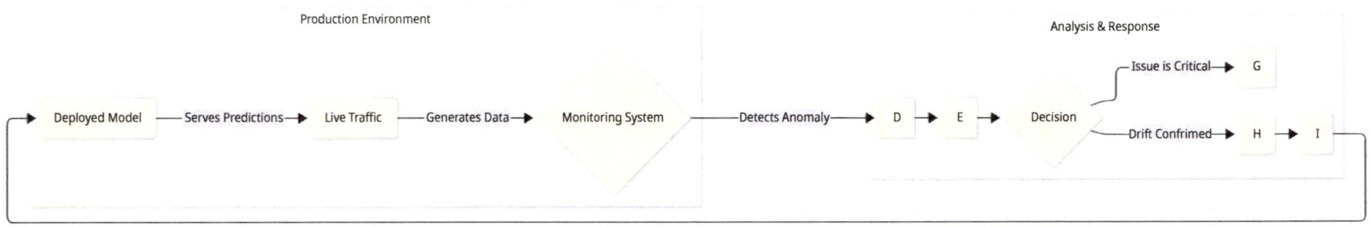

Figure 2.3: The MLOps governance loop. This diagram shows the continuous cycle of monitoring for anomalies in a production model, diagnosing the cause, and triggering an appropriate response, such as a rollback or a model retrain, to maintain system health. Source: Authors' own work.

triage notes to predict patient disposition. A 2024 study, for example, used a combination of NLP and traditional ML to predict the need for ICU admission based on initial triage information [49, 50].

Methods: The model ingested structured patient data alongside free-text triage notes. The text was processed using NLP techniques to extract features, which were then combined with the structured data and fed into an ensemble model, such as a random forest or GBM [49]. The inclusion of text from nurse notes was found to significantly enhance the model's predictive performance, as the notes often contained crucial context and symptoms not captured in discrete data fields [50].

Evaluation and Outcomes: The models were evaluated using metrics appropriate for the high-stakes medical context, including F_1-score to balance precision and recall, and the Brier score to assess probability calibration. A well-calibrated probability is essential for clinicians to trust the model's risk score. The study found that the AI model could outperform human physicians in predicting severe outcomes, demonstrating its potential as a powerful clinical decision support tool [49]. The decision threshold of the model was carefully adjusted to prioritize high sensitivity (recall), reflecting the clinical reality that a false negative (missing a critically ill patient) is far more costly than a false positive (over-triaging a stable patient) [50].

Challenges: Key challenges include the "messiness" of clinical text (abbreviations and typos), which requires robust NLP preprocessing. Furthermore, ensuring the model generalizes across different hospitals with varying patient populations (a form of distribution shift) is critical and requires external validation. Finally, gaining the trust of clinicians is paramount, which underscores the need for model explainability (e.g., using techniques like Local Interpretable Model-agnostic Explanations (LIME) to highlight which words in the notes most influenced the prediction) and a human-in-the-loop design [51].

2.9.2 Finance: Market Prediction with News Sentiment Analysis

Problem: Financial markets are notoriously complex and are influenced by a torrent of information, including news articles, regulatory filings, and social media commentary. Traders and investment firms seek to gain an edge by processing this information faster and more accurately than the market to predict stock price movements.

Solution: NLP-powered sentiment analysis models are widely used to quantify the emotional tone of financial text, transforming unstructured news and social media data into a predictive feature for quantitative trading models [52].

Methods: A typical approach involves using a domain-specific, pretrained language model like FinBERT (a BERT model fine-tuned on financial text) to analyze news headlines or articles and classify their sentiment as positive, negative, or neutral [53]. This sentiment score is then used as an input feature in a time-series model that predicts the next day's stock price direction or volatility. Recent studies show that modern LLMs can further improve the nuance of sentiment classification, distinguishing, for instance, between "earnings fell" (negative) and "earnings fell but beat expectations" (often market-positive) [54].

Evaluation and Outcomes: The success of these models is measured not just by classification accuracy but by their financial impact. This is typically assessed by back-testing a trading strategy that uses the model's sentiment signals. Studies consistently show that incorporating textual sentiment improves the directional accuracy of stock market prediction models and can lead to portfolios with higher returns or better risk-adjusted performance (Sharpe ratio) compared to baselines that use only price data [52].

Challenges: Financial language is highly nuanced, context-dependent, and rife with domain-specific jargon, making accurate sentiment classification difficult [55]. The market is also a highly adversarial environment, where models must be robust to "fake news" or coordinated social media campaigns designed to manipulate stock prices. Most importantly, financial markets are subject to constant and abrupt regime shifts (a form of concept drift), meaning models must be continuously monitored and retrained on recent data to remain effective [52].

2.9.3 Customer Service: Automated Support Ticket Routing and Prioritization

Problem: Large enterprises receive thousands of customer support tickets daily via email, web forms, and chat. Manually triaging and routing these tickets to the correct department (e.g., billing, technical support, and sales) is slow, inefficient, and can lead to poor customer experiences.

Solution: NLP-based text classification models can be deployed to automatically read incoming support tickets and predict the appropriate category, priority, and even the customer's sentiment (e.g., angry and frustrated) [56].

Methods: This is a classic multi-class text classification problem. Modern systems often use a fine-tuned transformer model (like BERT) to encode the subject and body of a support ticket into a vector representation. This vector is then fed into a classification layer that predicts the correct routing destination [56]. The model learns to associate specific keywords, phrases, and contexts with different support categories, going beyond simple keyword matching. For example, it can learn that a ticket containing "camera not working after update" belongs to "Hardware Issues" even if the word "hardware" is never mentioned [57].

Evaluation and Outcomes: The primary metric is *routing accuracy* – the percentage of tickets sent to the correct team. Secondary business metrics include reductions in average ticket handling time and improvements in customer satisfaction scores, which are direct results of faster and more accurate routing. Companies deploying these systems report high accuracy (often over 90%) and significant operational efficiencies, freeing up human agents from manual triage to focus on solving customer problems [58].

Challenges: A key challenge is the sheer variety of ways customers describe their problems, including nontechnical language, typos, and multiple languages. To manage the risk of misclassification, many systems operate with a confidence threshold: if the model's prediction confidence is below a certain level, the ticket is flagged for manual human review. The system must also be designed to adapt to *concept drift*, as the definitions of categories can change when new products are launched or organizational structures are changed. This necessitates a tight MLOps loop for continuous monitoring and retraining [69].

2.10 The Future of Predictive Modeling

The field of predictive modeling is evolving at an unprecedented pace, driven by advances in model architectures, computational scale, and the increasing availability of diverse data. Several key trends are shaping the next generation of predictive systems, promising models that are more powerful, efficient, trustworthy, and integrated into the fabric of decision-making.

2.10.1 Beyond a Single Modality: The Rise of Multimodal AI

The future of AI is *multimodal*. While many current systems operate on a single type of data (unimodal), future models will increasingly learn from and reason over multiple data types – such as text, images, audio, and structured data – simultaneously [59]. Humans naturally integrate information from multiple senses to understand the world; for example, a doctor reads a patient's clinical notes (text), examines an X-ray (image), and listens to their breathing (audio) to make a diagnosis. AI is moving in the same direction.

Recent large-scale models like Google's Gemini and OpenAI's GPT-4 are natively multimodal, capable of processing and connecting information across text, images, and audio in a single, unified architecture [12]. This will enable a new class of predictive applications. A predictive maintenance model could analyze both numerical sensor data and the sound of a machine to predict failure with greater accuracy. A customer churn model could process a user's support ticket text, their in-app clickstream data, and a recording of their support call to generate a more holistic risk score. While challenges remain in fusing heterogeneous data and managing the complexity of these models, the trend toward unified, multimodal reasoning is clear and will significantly expand the scope and power of predictive modeling [60].

2.10.2 Grounding Predictions in Reality: Retrieval-Augmented Generation (RAG)

A key limitation of traditional LLMs is that their knowledge is "frozen" at the time of training. They can become outdated and are prone to "hallucinating" or fabricating facts. *RAG* is an architectural pattern that addresses this by combining a pretrained generative model with an external knowledge retrieval mechanism [61].

In a RAG system, when a query is received, the system first retrieves relevant documents or data snippets from an up-to-date external knowledge base (e.g., a corporate database, a collection of recent news articles, or the live web). This retrieved information is then provided as context to the language model along with the original query. The model uses this fresh, factual context to generate its response, effectively grounding its prediction in external reality [62]. RAG reduces hallucinations, improves the factual accuracy of predictions, and allows models to provide citations for their outputs, enhancing transparency and trust [63]. This approach is transforming predictive applications that rely on timely information, from financial question-answering systems that retrieve real-time market data to enterprise chatbots that access the latest internal documentation.

2.10.3 The Push for Efficiency: Knowledge Distillation and Quantization

While massive, billion-parameter models dominate research headlines, a parallel and equally important trend is the development of small, efficient models suitable for deployment on resource-constrained devices like smartphones, wearables, and Internet of Things sensors. This field, sometimes called TinyML, is driven by the need for low latency, low power consumption, and enhanced privacy (as data can be processed locally on-device).

Two key techniques for creating efficient models are:

- **Knowledge Distillation**: This is a model compression technique where the knowledge from a large, powerful "teacher" model is transferred to a smaller, faster "student" model [64]. The student model is trained not just on the ground-truth labels but also to mimic the output probability distributions (the "soft targets") of the teacher model. By learning from the teacher's nuanced predictions, the student can often achieve performance close to that of the much larger teacher model, but at a fraction of the computational cost [65].
- **Quantization**: This technique reduces the numerical precision of a model's weights and activations, for example, by converting them from 32-bit floating-point numbers to 8-bit integers. This can lead to a significant reduction in model size (up to 4×) and a substantial increase in inference speed, especially on hard-

ware with specialized support for integer arithmetic, often with only a minimal drop in accuracy [66].

Together, these and other efficiency-focused innovations are enabling the deployment of sophisticated predictive models in environments previously thought impossible, from real-time voice recognition on smartwatches to predictive health monitoring on wearable sensors.

2.11 Conclusion

Predictive modeling with AI has evolved from a niche academic discipline into a transformative force reshaping industries. By leveraging historical data, powerful algorithms, and rigorous engineering practices, organizations can build systems that forecast outcomes with remarkable accuracy, driving efficiency, innovation, and competitive advantage. This chapter has charted the comprehensive journey of predictive modeling, from its foundational principles to its deployment in the real world and its future trajectory.

We have seen that while sophisticated algorithms like the transformer have unlocked new frontiers, particularly in the realm of NLP, enduring success hinges on a disciplined, end-to-end approach. The quality of a model is inextricably linked to the quality of its data pipeline. The reliability of its predictions depends on a meticulous and context-aware evaluation strategy. And its long-term value is contingent on a robust MLOps framework that can manage the inevitable challenges of a changing world, such as distribution drift and the emergence of bias.

As AI systems become more powerful and pervasive, the principles of responsible AI – robustness, fairness, privacy, and transparency – are no longer optional considerations but are central to building trust and ensuring sustainable adoption. The future of the field points toward models that are even more capable, integrating multiple modalities of data and grounding their predictions in real-time knowledge. Simultaneously, a parallel drive for efficiency is making these powerful capabilities more accessible, pushing predictive intelligence to the edge and into the devices that permeate our daily lives.

To navigate this complex but promising landscape, practitioners must blend technical expertise with a deep understanding of the domain, a commitment to ethical principles, and a disciplined engineering mindset. The following checklist distills the key takeaways from this chapter into a practical guide for developing and deploying predictive models responsibly and effectively. By adhering to these principles, practitioners can harness the power of predictive AI to create significant real-world value while thoughtfully managing its risks.

2.12 A Practitioner's Checklist for Predictive Modeling

This checklist provides a concise summary of best practices for developing and deploying predictive models responsibly and effectively.

Phase 1: Problem Definition and Data Foundation

- **Define the Prediction Target**: Clearly state the business problem and the exact variable to be predicted
- **Set Success Criteria**: Define both a technical metric (e.g., PR-AUC > 0.8) and a business KPI (e.g., 10% reduction in churn)
- **Understand Data Provenance**: Document data sources, collection methods, and any known limitations or biases
- **Perform Exploratory Data Analysis (EDA)**: Analyze feature distributions, correlations, and data quality issues
- **Identify Potential Biases**: Proactively identify sensitive attributes and potential proxy variables in the data

Phase 2: Development and Validation

- **Establish a Proper Validation Strategy**: Use a time-based split for temporal data or a stratified split for imbalanced classification. Keep a hold-out test set pristine until final evaluation
- **Build a Robust Feature Pipeline**: Ensure all preprocessing steps (scaling, imputation) are fitted *only* on the training data to prevent leakage
- **Prevent Target Leakage**: Scrutinize every feature to confirm it would be available at prediction time
- **Start with a Simple Baseline**: Train a simple, interpretable model (e.g., logistic regression) to establish a performance baseline
- **Select Appropriate Metrics**: For imbalanced problems, prioritize PR-AUC, F_1-score, or a cost-sensitive metric over accuracy
- **Tune Hyperparameters Systematically**: Use cross-validation with a systematic search method (e.g., random search and Bayesian optimization)
- **Check Model Calibration**: Evaluate if the model's predicted probabilities are reliable. If not, apply a calibration technique like temperature scaling
- **Evaluate on Subgroups**: Analyze model performance across different demographic or business segments to detect fairness issues
- **Document the Model**: Create a Model Card detailing intended use, training data, evaluation results (including fairness metrics), and limitations [41]

Phase 3: Deployment and Operations (MLOps)

- **Version Everything**: Use version control for data, code, and model artifacts to ensure reproducibility

- **Automate the Pipeline**: Implement a CI/CD pipeline to automate testing, validation, and deployment
- **Prevent Training-Serving Skew**: Use a feature store or ensure identical feature logic in training and serving environments [68]
- **Implement a Monitoring Plan**: Set up monitors for:
  - Operational health (latency and errors)
  - Data and concept drift
  - Predictive performance (via a feedback loop)
  - Fairness and bias metrics
- **Use a Safe Deployment Strategy**: Employ canary, A/B, or shadow testing for new model rollouts
- **Establish a Rollback Plan**: Ensure you can quickly revert to a previous stable model version if an issue is detected
- **Schedule Periodic Retraining**: Define a strategy for regularly updating the model on fresh data to combat drift
- **Secure the System**: Protect model APIs from abuse and ensure compliance with data privacy regulations (e.g., GDPR)
- **Incorporate a Human-in-the-Loop**: For high-stakes decisions, define criteria for escalating low-confidence predictions or edge cases to a human expert for review

References

[1] IBM. (2022). *What Is Predictive Analytics?*. IBM.

[2] Authors, V. (2025). *The Foundations of Machine Learning and Deep Learning*.

[3] Bishop, C. M. (2006). *Pattern Recognition and Machine Learning*. Springer.

[4] Holdsworth, J., & Scapicchio, M. (2024). *What Is Deep Learning?*. IBM.

[5] Belle, V., & Papantonis, I. (2021). Principles and practice of explainable artificial intelligence. *Frontiers in Big Data*, 4. https://doi.org/10.3389/fdata.2021.688969

[6] Kutz, J. (2025). *Tokenization Vs Embedding How are They Different?*. Airbyte.

[7] Devlin, J., Chang, M.-W., Lee, K., & Toutanova, K. (2019). BERT: Pre-training of deep bidirectional transformers for language understanding. *Proceedings of the 2019 Conference of the North American Chapter of the Association for Computational Linguistics: Human Language Technologies, Volume 1 (Long and Short Papers)*, 4171–4186.

[8] Vaswani, A., Shazeer, N., Parmar, N., Uszkoreit, J., Jones, L., Gomez, A. N., Kaiser, Ł., & Polosukhin, I. (2017). Attention is all you need. *Advances in Neural Information Processing Systems*, 30.

[9] Hivenet. (2025). *A Beginner's Guide to the Transformer Architecture in Deep Learning*.

[10] Polo Club of Data Science. (2025). *LLM Transformer Model Visually Explained*.

[11] Nightfall AI. (2025). *Transformer architectures: The Essential Guide*.

[12] OpenAI. (2023). GPT-4 technical report. *arXiv Preprint arXiv:2303.08774*.

[13] Sculley, D., Holt, G., Golovin, D., Davydov, E., Phillips, T., Ebner, D., Chaudhary, V., Young, M., Crespo, J.-F., & Dennison, D. (2015). Hidden technical debt in machine learning systems. *Advances in Neural Information Processing Systems*, 28.

[14] IBM. (2025). *What Is Data Leakage in Machine Learning?*.

[15] Cook, A. (2025). *Data Leakage*.

[16] Regunath, G. (2025). *A Beginner's Guide to Understanding Feature Stores.*
[17] Anyscale. (2025). *Feature Store – Made with ML.*
[18] Wolpert, D. H., & Macready, W. G. (1997). No free lunch theorems for optimization. *IEEE Transactions on Evolutionary Computation, 1*(1), 67–82.
[19] Breiman, L. (2001). Random forests. *Machine Learning, 45*(1), 5–32.
[20] Chen, T., & Guestrin, C. (2016). Xgboost: A scalable tree boosting system. *Proceedings of the 22nd Acm Sigkdd International Conference on Knowledge Discovery and Data Mining*, 785–794.
[21] LeCun, Y., Bottou, L., Bengio, Y., & Haffner, P. (1998). Gradient-based learning applied to document recognition. *Proceedings of the IEEE, 86*(11), 2278–2324.
[22] Gu, A., Goel, K., & Ré, C. (2022). Efficiently modeling long sequences with structured state spaces. *International Conference on Learning Representations (ICLR).* https://arxiv.org/abs/2111.00396
[23] Google for Developers. (2025). *Classification: ROC and AUC.*
[24] Saito, T., & Rehmsmeier, M. (2015). The precision-recall plot is more informative than the ROC plot when evaluating binary classifiers on imbalanced datasets. *PLOS ONE, 10*(3). e0118432. https://doi.org/10.1371/journal.pone.0118432.
[25] Brownlee, J. (2025b). *ROC AUC Vs Precision-recall for Imbalanced Data.*
[26] Daily Dose of Data Science. (2025). *A Crash Course of Model Calibration – Part 1.*
[27] Guo, C., Pleiss, G., Sun, Y., & Weinberger, K. Q. (2017). On calibration of modern neural networks. *Proceedings of the 34th International Conference on Machine Learning, 70*, 1321–1330.
[28] Suleman, E. (2025). *A Gentle Guide to Model Calibration.*
[29] Platt, J. (1999). Probabilistic outputs for support vector machines and comparisons to regularized likelihood methods. *Advances in Large Margin Classifiers*, 61–74.
[30] Zadrozny, B., & Elkan, C. (2002). Transforming classifier scores into accurate multiclass probability estimates. *Proceedings of the Eighth ACM SIGKDD International Conference on Knowledge Discovery and Data Mining*, 694–699.
[31] Brownlee, J. (2025a). *Cost-sensitive Learning for Imbalanced Classification.*
[32] Fraud Detection Handbook. (2025). *Cost-sensitive Learning.*
[33] GeeksforGeeks. (2025a). *How to Implement Cost-sensitive Learning in Decision Trees.*
[34] Kulinski, S. (2025). *Distribution Shifts and How We Can Learn from Them.*
[35] NeurIPS. (2021). *Distribution Shifts: Connecting Methods and Applications (Distshift).*
[36] Zhang, A., Lipton, Z. C., Li, M., & Smola, A. J. (2021). *Environment and Distribution Shift.*
[37] Evidently AI. (2025). *What Is Data Drift in ML, and How to Detect and Handle It.*
[38] Mehrabi, N., Morstatter, F., Saxena, N., Lerman, K., & Galstyan, A. (2021). A survey on bias and fairness in machine learning. *ACM Computing Surveys, 54*(6). 1–35. https://doi.org/10.1145/3457607.
[39] Verma, S., & Rubin, J. (2018). Fairness definitions explained. *Proceedings of the 2018 IEEE/ACM International Workshop on Software Fairness*, 1–7.
[40] Zhang, B. H., Lemoine, B., & Mitchell, M. (2018). Mitigating unwanted biases with adversarial learning. *Proceedings of the 2018 AAAI/ACM Conference on AI, Ethics, and Society*, 335–340. https://doi.org/10.1145/3278721.3278779
[41] Mitchell, M., Wu, S., Zaldivar, A., Barnes, P., Vasserman, L., Hutchinson, B., Spitzer, E., Raji, I. D., & Gebru, T. (2019). Model cards for model reporting. *Proceedings of the Conference on Fairness, Accountability, and Transparency*, 220–229.
[42] Shokri, R., Stronati, M., Song, C., & Shmatikov, V. (2017). Membership inference attacks against machine learning models. *2017 IEEE Symposium on Security and Privacy (SP)*, 3–18.
[43] GeeksforGeeks. (2025b). *Privacy-preserving in Machine Learning.*
[44] Abadi, M., Chu, A., Goodfellow, I., McMahan, H. B., Mironov, I., Talwar, K., & Zhang, L. (2016). Deep learning with differential privacy. *Proceedings of the 2016 ACM SIGSAC Conference on Computer and Communications Security*, 308–318. https://doi.org/10.1145/2976749.2978318

[45] Konečný, J., McMahan, H. B., Yu, F. X., Richtárik, P., Suresh, A. T., & Bacon, D. (2016). Federated learning: Strategies for improving communication efficiency. *arXiv Preprint arXiv:1610.05492*.

[46] Superblocks. (2023). *What Is AI Model Governance? Why It Matters & Best Practices*. Superblocks Blog.

[47] Amazon Web Services. (2025). *What Is MLOps? – Machine Learning Operations Explained*.

[48] MLOps.org. (2025). *MLOps Principles*.

[49] Chang, Y.-H., Lin, Y.-C., Huang, F.-W., Chen, D.-M., Chung, Y.-T., & Chen, W.-K. (2024). Using machine learning and natural language processing in triage for prediction of clinical disposition in the emergency department. *BMC Emergency Medicine*, *24*(1). 237. https://doi.org/10.1186/s12873-024-01152-1.

[50] Kaufmann, M., & Rohlfing, K. J. (2024). Explainable NLP model for predicting patient admissions at emergency department using triage notes. *Proceedings of the 15th International Conference on Agents and Artificial Intelligence*.

[51] ResearchGate. (2024). *(PDF) Explainable NLP Model for Predicting Patient Admissions at Emergency Department Using Triage Notes*.

[52] Darwish, M., Hassanien, A. E., & Eissa, A. H. (2025). Stock market forecasting: From traditional predictive models to large language models. *Computational Economics*. https://doi.org/10.1007/s10614-025-11024-w

[53] Araci, D. (2019). Finbert: Financial sentiment analysis with pre-trained language models. *arXiv Preprint arXiv:1908.10063*.

[54] Liu, W., Chen, Y., Zhang, Z., & Zhang, Q. (2025). *Revisiting Financial Sentiment Analysis: A Language Model Approach*.

[55] Loughran, T., & McDonald, B. (2011). When is a liability not a liability? Textual analysis, dictionaries, and 10-ks. *The Journal of Finance*, *66*(1), 35–65.

[56] Ndlovu, F. (2023). *Intelligent Support Ticket Routing with NLP*. DEV Community.

[57] SentiSum. (2025). *How to Automate Ticket Classification – NLP for Support Tickets*.

[58] EverWorker. (2025). *AI Ticket Prioritization and Routing: A Complete Guide*.

[59] Li, K., et al. (2024). *Multimodal Deep Learning: A Survey of Models, Fusion Strategies, Applications and Research Challenges*.

[60] Baltrušaitis, T., Ahuja, C., & Morency, L.-P. (2018). Multimodal machine learning: A survey and taxonomy. *IEEE Transactions on Pattern Analysis and Machine Intelligence*, *41*(2), 423–443.

[61] Lewis, P., Perez, E., Piktus, A., Petroni, F., Karpukhin, V., Goyal, N., Küttler, H., Lewis, M., Yih, W., Rocktäschel, T., Riedel, S., & Kiela, D. (2020). Retrieval-augmented generation for knowledge-intensive NLP tasks. *Advances in Neural Information Processing Systems*, *33*, 9459–9474.

[62] Merritt, R. (2025). *What Is Retrieval-augmented Generation, Aka RAG?*. NVIDIA Blog.

[63] SingleStore. (2025). *RAG Tutorial: A Beginner's Guide to Retrieval Augmented Generation*.

[64] Hinton, G., Vinyals, O., & Dean, J. (2015). Distilling the knowledge in a neural network. *arXiv Preprint arXiv:1503.02531*.

[65] neptune.ai. (2025). *Knowledge Distillation: Principles, Algorithms, Applications*.

[66] TensorFlow. (2025). *Quantization Aware Training Comprehensive Guide*.

[67] Aerospike. *Feature Store 101: Build, Serve, and Scale ML Features*. Aerospike Blog.

[68] Amazon Web Services. *MLREL-07: Ensure Feature Consistency across Training and Inference*. AWS Well-Architected Framework.

[69] Google Cloud. *MLOps: Continuous Delivery and Automation Pipelines in Machine Learning*. Google Cloud Architecture Center.

[70] Microsoft. *Databricks Feature Store*. Azure Databricks Documentation.

Kamesh Sonti, T. V. N. L. Aswini, T. J. V. Subrahmanyeswara Rao, Purnima K. Sharma

Chapter 3
Glacier Disappearance Prediction: Observational Constraints and Modeling Approaches

Abstract: The loss of glaciers is one of the most obvious and palpable effects of human-caused climate change. Globally, mountain glaciers are receding more quickly, which has consequences for alpine risks, water supplies, and sea level rise. The present knowledge of glacier mass balance, observational monitoring methods, and prediction techniques – from process-based, physically robust models to new machine learning approaches – are summarized in this chapter. We study regional case studies, go over global estimates of glacier loss under several warming scenarios, and talk about the consequences for policy and adaptation. Up to 50% of the world's glaciers could vanish with a 1.5 °C increase, according to recent satellite data and sophisticated modeling frameworks, underscoring the need for precise forecasts and prompt climate action.

Keywords: Glaciers, machine learning, satellite data, modeling frameworks

3.1 Introduction

One of the most sensitive markers of climate change, glaciers are trustworthy environmental sentinels for tracking changes in local and worldwide climate trends [1]. Rising temperatures, changing precipitation patterns, and longer melt seasons have caused extensive mass loss in almost all glacier regions since the 1990s [2]. Even under hopeful low-emission scenarios, the Intergovernmental Panel on Climate Change (IPCC) has demonstrated that glacial retreat and loss will remain a defining feature of the twenty-first century [1, 2].

Kamesh Sonti, Sri Vasavi Engineering College, Pedatadepalli 534101, Andhra Pradesh, India, e-mail: kamesh.sonti@gmail.com
T. V. N. L. Aswini, Sri Vasavi Engineering College, Pedatadepalli 534101, Andhra Pradesh, India, e-mail: aswinivnl@gmail.com
T. J. V. Subrahmanyeswara Rao, Sasi Institute of Technology and Engineering, Tadepalligudem 534101, Andhra Pradesh, India, e-mail: tjvsrao@gmail.com
Purnima K. Sharma, Chitkara University Institute of Engineering and Technology, Chitkara University, Rajpura 140417, Punjab, India, e-mail: purnima.sharma@chitkara.edu.in

 | https://doi.org/10.1515/9783112211991-003

Beyond glaciology, the rapid melting of alpine ice has significant ramifications. Hundreds of millions of people are impacted by glaciers, which control seasonal water supply in numerous heavily populated river basins throughout Central Asia, the Andes, and High-Mountain Asia [3]. Hazards like glacial lake outburst floods (GLOFs), avalanches from unstable slopes, and enhanced debris-flow activity are further made exacerbated by rapidly receding glaciers [4]. Globally, the Greenland Ice Sheet's contribution to sea level rise is currently less than that of glacier mass loss, and this contribution is increasing.

Robust prediction frameworks are needed to quantify the timing of glacier disappearance and the spatial evolution of this process across mountain ranges. These frameworks need to combine increasingly complex process-based and data-driven models with a variety of observational streams, such as field measurements and satellite remote sensing. The current state of glacier disappearance prediction is reviewed in this chapter, along with observational limitations, modeling techniques ranging from conventional energy-balance simulations to state-of-the-art machine learning (ML), global and regional forecasts, and new research avenues.

3.2 Glacier Mass Balance Fundamentals

The surface mass balance, which is defined as the net accumulation (snowfall, rime, and avalanche input) minus ablation (melt, sublimation, and wind erosion), is what essentially controls glacier evolution. A glacier must geometrically adjust by decreasing in size and thickness when it encounters persistent negative mass balance. The glacier may recede to a lesser equilibrium size or completely vanish if the negative balance continues [5].

Glacier health is strongly correlated with the accumulation area ratio (AAR), which is the percentage of a glacier's surface in the accumulation zone (where annual snow accumulation exceeds ablation). As a sign of increased mass loss and possible demise, glaciers with declining AARs indicate that less of the glacier surface is in the accumulation zone [5, 6].

Three complementary observational methods can be used to characterize mass balance: (a) in situ measurements using stakes, snow pits, and ablation markers; (b) geodetic methods comparing digital elevation models (DEMs) from different epochs; and (c) glaciological balance integrating field observations over a particular balance year [7]. Although each method has advantages and disadvantages, taken as a whole, they offer the empirical basis for validating and calibrating disappearance models.

3.3 Observational Monitoring and Satellite Remote Sensing

Since satellite technology became available in the 1970s, the systematic monitoring of glaciers has experienced a radical upheaval. A constellation of specialized cryospheric missions that provide previously unheard-of temporal and geographical resolution has replaced early optical satellites like Landsat [8].

3.3.1 Satellite-Derived Glacier Products

Periodic mapping of glacier size and surface reflectance is made possible by multispectral optical sensors (Landsat, Sentinel-2, ASTER). By using radar interferometry to assess ice velocity through cloud penetration, synthetic aperture radar (SAR) sensors (Sentinel-1) reveal glacier flow dynamics that control calving and dynamic imbalance [8, 9]. DEM differencing at sub-100 m resolution across vast areas is made possible by recent high-resolution commercial satellites and airborne LiDAR, providing accurate estimates of glacier-wide elevation and volume change [8].

3.3.2 Global Monitoring Initiatives

The most extensive dataset of measurements of glacier length, area, and mass balance is kept up to date by the World Glacier Monitoring Service (WGMS) [10]. About 215,000 glaciers and ice caps worldwide are listed in satellite-derived glacier inventories (such as the Randolph Glacier Inventory, RGI, and its successor, RGI 7.0) with uniform geometric parameters [11]. Detailed maps of glacier-wide mass change across multi-decadal periods are now available thanks to community-driven evaluations that combine in situ observations, satellite data, and statistical techniques. Published estimates show that glacial mass loss has in fact increased [12].

3.4 Process-Based Glacier Evolution Models

The physical basis for predicting glacier loss is represented by numerical models of glacier mass balance and flow. These models function over a range of computing costs and levels of complexity.

3.4.1 Temperature-Index Models

Temperature-index (degree-day) approaches, in which daily or monthly melt is proportional to positive temperatures, are the most basic and widely used methodologies [13]. Temperature-index models can be coupled to global climate models at regional and global scales because they only need temperature and precipitation fields as input [13]. When calibrated against observed mass balance, temperature-index models have demonstrated efficacy for multi-decadal projections despite their simplicity.

3.4.2 Full Energy-Balance Models

To calculate surface energy balance, more physically explicit models take into consideration sensible heat, latent heat, precipitation, solar radiation, and atmospheric longwave radiation [13]. When simulating complicated glaciers covered in debris and high-elevation plateaus where temperature changes alone are insufficient, energy-balance models perform better [13]. However, their applicability is limited to small research locations or carefully chosen benchmark glaciers due to their data needs and computational cost.

3.4.3 Glacier Flowline and 3D Models

Glacier geometry change, caused by both mass balance and ice dynamics can be predicted using dynamic ice-flow models that solve for ice thickness and velocity [14]. While mass balance predominates and more straightforward methods are adequate for many small alpine glaciers, accounting for ice flow is crucial for surge-type glaciers and quickly calving outlets [14]. Although they are still computationally costly, full 3D models are becoming increasingly popular for high-resolution projections of massive ice masses.

3.4.4 Global Glacier Evolution Models

Large-scale frameworks designed to simulate all glaciers (excluding ice sheets) include:

- **OGGM (Open Global Glacier Model):** Each glacier is represented as a flowline with specified ice dynamics using a modular, open-source framework [15]. Under CMIP5 and CMIP6 climate forecasts, glacier change has been simulated using OGGM.

- **GloGEM (Global Glacier Evolution Model):** Using spatial mass-balance parameterizations, this glacier-centric model is scalable from individual glaciers to regions [16].
- **PyGEM (Python Glacier Evolution Model):** A Python-based framework that allows for variable representation of ice dynamics, mass balance, and calving in conjunction with projections from climate models [17].
- Under both low- and high-emission scenarios, these frameworks reliably predict that glaciers will lose 18–36% of their early twenty-first-century mass by 2100, with significant regional variation [1, 2]. Crucially, at the lower end of uncertainty ranges, recent observed mass loss has matched or surpassed the rates predicted by these models, indicating that certain disappearance timescales may have been conservative.

3.5 Statistical and ML Approaches to Mass Balance Prediction

A complementary set of data-driven and ML techniques for estimating and projecting glacier mass balance have been made possible due to the growing availability of glacier and climatic measurements.

3.5.1 Statistical Regression Models

Climate indices (such as summer temperature, winter precipitation, and the North Atlantic Oscillation) or gridded climate fields are directly linked to observed mass balance using empirical regression models [18]. Although these models may capture nonlinear interactions and are computationally quick, their transferability across regions with significantly different underlying climate-mass balance relationships is frequently restricted [18].

3.5.2 Random Forest and Ensemble Methods

Glacier mass-balance measurements have been used to train ensemble ML techniques, including random forests and gradient boosting, to identify nonlinear correlations and optimal feature importance [19]. When tested against separate test sets, pilot investigations show that these approaches can outperform conventional temperature-index models in terms of prediction error [19].

3.5.3 Seasonal and Annual Prediction

Recent extensive research has shown that sophisticated ML models, trained on hemispheric or global mass balance datasets, perform better than traditional temperature-index models in forecasting the annual and seasonal mass balance of glaciers in a variety of geographical locations [20]. The potential of data-driven approaches for generalized glacier prediction was highlighted by a meta-analysis that compared ML mass balance models to three cutting-edge glacier evolution models (such as OGGM and GloGEM) and found that the ML approach achieved lower root-mean-square error and bias when evaluated against both in situ and geodetic observations [20].

3.5.4 2024–2025 Machine Learning Benchmarking

In a groundbreaking 2025 study published in *The Cryosphere* [20] which are shown in Table 3.1, three well-known glacier evolution models (GloGEM, OGGM, and PyGEM) were directly compared to a ML mass balance model (MBM) on independent test glaciers. When compared to the best-performing glacier evolution model, MBM reduced root-mean-square error on seasonal mass balance by up to 46% for winter and 25% for summer, with minimal bias (−0.05 to +0.10 m w.e.) as opposed to significant biases (+0.26 to +0.59 m w.e.) in conventional models [20]. This study provides empirical evidence that ML can reliably generalize across a range of glaciers and climatic conditions, opening up new possibilities for rapid and cost-effective mass balance prediction on glacier locations with insufficient monitoring [20].

Table 3.1: Model performance comparison on seasonal and annual glacier mass balance (2025 benchmarking study).

Model	Winter RMSE (m w.e.)	Summer RMSE (m w.e.)	Annual bias (m w.e.)	Computational cost
Machine learning (MBM)	0.54	0.59	−0.01	Very low
OGGM (temperature index)	0.71	0.69	+0.02	Moderate
GloGEM (spatial MB)	0.78	0.72	−0.03	High
PyGEM (flexible)	0.81	0.75	+0.04	High

3.5.5 Hybrid and Physics-Informed Approaches

Using physics-informed neural networks (PINNs) and hybrid architectures, a new field integrates ML with process-based understanding [21]. In order to guarantee that taught models adhere to basic physical principles while utilizing data to capture intricate sub-grid activity, these methods incorporate conservation laws (energy balance, mass balance, momentum equations) as regularization constraints within neural network training [21]. Further performance improvements have been attained in recent hybrid studies that combine ML with traditional glacio-hydrological models, especially for complicated debris-covered and surge-type glacier systems [36].

3.6 Deep Learning and Multimodal Frameworks

New opportunities for glacier observation and prediction have been made possible by advancements in deep learning and computer vision.

3.6.1 Convolutional Neural Networks for Glacier Extent

Convolutional neural networks (CNNs) trained on optical satellite images can quickly evaluate historical picture archives and reduce manual labor by autonomously segmenting glacier limits with high accuracy [22]. CNNs may generate probabilistic forecasts of future extent and learn temporal patterns of glacier retreat when trained on multi-temporal picture stacks [22].

3.6.2 Multimodal Architectures

In order to infer glacier mass balance, ice velocity, and future evolution, recent frameworks combine optical images, SAR, DEMs, and climatic fields into unified deep learning models [23]. These architectures can capture intricate interactions between topography, climate forcing, surface features, and glacier response by learning representations from many data modalities [23].

3.6.3 Fast Emulators of High-Fidelity Models

Deep neural networks have been effectively trained as computationally quick replicas of costly models of glaciers and ice sheets [24]. Neural networks, for example, have executed orders of magnitude faster and repeated multi-year simulations of ice-shelf melt rates with near ocean-model accuracy [24]. Large-ensemble forecasts of mountain glacier loss can be accelerated by similar emulator approaches, allowing uncertainty quantification across millions of parameter variations.

3.7 Global Projections of Glacier Disappearance

A clear picture emerges from the synthesis of data and modeling: glaciers are melting, and as global warming continues, the pace of melting will increase.

3.7.1 Temperature Sensitivity

Every 0.5 °C increase in global temperature causes a nonlinear increase in glacier mass loss and disappearance, according to world-scale evaluations utilizing satellite-derived mass-change estimations and glacier evolution models [25]. The basic sensitivity of glacier disappearance timeframes to global temperature trajectories is shown by the fact that this link is consistent across all emission scenarios and climate models analyzed [25].

3.7.2 1.5 °C and 2 °C Warming Thresholds

About 50% of the world's glaciers (by number, mostly smaller ice bodies) are predicted to vanish under a 1.5 °C warming scenario compared to pre-industrial conditions [26]. This fraction increases significantly at 2 °C warming, with equivalent sea-level contributions of several centimeters in both scenarios [25, 26]. The number and spatial extent of disappearing glaciers rise almost linearly with further warming above 2 °C [25, 26].

3.7.3 2025 Updated Projections

The most recent glacier estimates, based on updated climate models and newly accessible satellite data through 2024, are presented in a 2025 study published in Science, as mentioned in Table 3.2 [36]. According to these revised estimates, a 1.5 °C warming limit would conserve 47% of the world's glacier mass by 2100 (compared to the 2020 baseline), contributing 138 mm to sea level increase, whereas a 2.0 °C warming limit would save only 37% of the glacier mass, raising sea levels by 190 mm [36]. When comparing 1.5 °C to 2.0 °C scenarios, these estimates show a nearly 40% decrease in

Table 3.2: Projected global glacier mass loss and sea-level contribution by temperature scenario (2025 assessment).

Temperature scenario	Glacier mass loss (%)	Mass remaining (%)	Sea-level rise (mm)	Glaciers disappearing (%)
1.5 °C	47–53	47 (20–64 range)	138 (59–237 range)	49
2.0 °C	63	37 (43–76 range)	190 (140–260 range)	62
3.0 °C	72	28	220	75
4.0 °C	79	21	245	83

glacier preservation, highlighting how crucial it is to meet strict emission objectives [35].

3.7.4 Regional Variation

The retreat of glaciers is not evenly distributed. With >80% mass loss predicted by 2100 even under low-emission scenarios, small glaciers in semi-arid locations (parts of the Andes, Central Asia, and the Mediterranean) are most at risk of extinction [1, 27]. Large ice masses in Greenland, Antarctica, and the main High-Asian glaciers, on the other hand, will endure longer while having drastically decreased thickness and area [1, 27]. Glaciers in Western Canada and the contiguous USA are already suffering accelerated losses that surpass previous IPCC forecasts, according to recent 2024 assessments [37].

3.7.5 Sea-Level Contribution

By 2100, the cumulative sea-level rise from melting glaciers and ice caps is expected to surpass 0.1 m, which is equal to Greenland's expected contributions and competes with thermal expansion as a cause of coastal flooding [2, 25]. Glacier preservation is one of the most economical climate mitigation goals, as updated 2025 projections indicate that glacier contributions may reach 0.19–0.24 m under moderate to high warming scenarios [35].

3.8 Regional Case Studies and Implications

3.8.1 High-Mountain Asia

The variability of glacial reaction is exemplified by High-Mountain Asia, which has the largest glacier reserve outside the poles in the Second Pole [28]. The Himalayan and Hindu Kush glaciers show complicated patterns: most of them disappear quickly, while other areas (like Karakoram) have seen unusual stability or modest growth [28]. Mass balance estimates for Northwestern Himalayan benchmark glaciers were updated in 2025 using high-resolution Sentinel-1A SAR, confirming accelerated losses with 8 year observations (2016–2024) demonstrating sustained negative mass balance across all surveyed sites [32]. Consensus indicates significant mass loss and the disappearance of the majority of minor glaciers by 2100, although disappearance forecasts are highly susceptible to local aerosol impacts and regional climate models [28].

3.8.2 European Alps

Since industrialization, Alpine glaciers have drastically receded, with an area loss of more than 50% since 1900 [29]. The Alps could lose more than 80% of their current glacier volume by 2100, even under strict emission scenarios; many tiny glaciers will have completely disappear in a matter of decades [29]. The seasonality of runoff, hydropower output, and water supply to downstream populations in Central and Western Europe are all significantly impacted by this trajectory [29].

3.8.3 Andes

About 70% of the world's mountain glaciers outside of High Mountain Asia are found in the tropical and subtropical Andes, but because of their smaller mass and equatorial location, they are especially susceptible to disappearing [30]. By 2100, the Andean glaciers in the central Andes are predicted to almost completely disappear, which would have a significant impact on the supply of water in Peru, Bolivia, and Chile [30].

3.8.4 Glacial Lake Outburst Hazards

Proglacial lakes grow and increase as glaciers recede, raising the possibility of catastrophic GLOFs that might destroy settlements further downstream [4]. By combining estimates of glacier extent with evaluations of lake stability and downstream exposure, ML, and automated hazard identification techniques are currently being used to anticipate which retreating glacier-lake systems present the most future risk [31].

3.9 Uncertainties, Limitations, and Research Gaps

Significant uncertainty remains in the prediction of glacier disintegration timescales and spatial patterns, despite considerable advances.

3.9.1 Structural Model Uncertainties

Numerous parameterizations (such as melt factors, albedo deterioration, and calving velocities) that vary locally and are frequently poorly defined are used in process-based models [32]. Projections from ensembles of glacier models subjected to the same climate scenario differ significantly, with mass change standard deviations frequently matching the ensemble mean [32].

3.9.2 Climate Forcing Uncertainties

Global climate models reveal large biases in reproducing regional temperature and precipitation at glacier-relevant scales [33]. Not all structural biases can be eliminated by statistical downscaling and bias-correction techniques, especially when it comes to precipitation, which is infamously hard to model in mountainous areas [33].

3.9.4 Data Sparsity and Observational Gaps

Opportunities to calibrate and validate models are limited since many distant glaciers lack in situ observations [10]. Although satellite-derived products have significantly improved, temporal gaps, cloud cover, and the challenge of detecting ice thickness continue to be major obstacles [8, 34].

3.9.5 Machine Learning Generalization

In remote, poorly sampled glaciers, ML-based methods developed on well-monitored glacier regions might not transfer consistently [19]. There are still issues with addressing regional bias in training data and making sure ML models can generalize to new climate regimes [19, 21].

3.9.6 Process Representation

There is still a lack of understanding and parameterization regarding debris cover, hanging glaciers, surge dynamics, and the relationship between glacier geometry and flow [5, 32]. On decadal periods, these mechanisms have the potential to significantly change disappearance timelines [32].

3.10 Future Directions and Outlook

3.10.1 Advancing Observational Networks

Richer restrictions on glacier evolution will be provided by growing satellite constellations with shorter return times and higher resolution [8]. To calibrate and evaluate models in a variety of glaciological situations, investments in in situ networks in under-sampled regions (such as Central Asia and the outer Himalayas) are crucial [10].

3.10.2 Physics-Informed ML

While maintaining the interpretability and computational efficiency benefits of ML, hybrid approaches that combine physical conservation rules with data-driven learning hold promise for lowering structural model uncertainty [21]. Methodological advancements will be accelerated by the creation of defined benchmarks and validation procedures for contrasting process-based, statistical, and ML techniques [20].

3.10.3 Integrated Hazard and Impact Modeling

More practical predictions for adaptation planning will be made possible by connecting projections of glacier retreat to downstream effects, such as seasonal water supply, GLOF risk, and ecosystem change [4, 28]. In order to guarantee that scientific projections meet actual societal demands, the co-production of information with stakeholders in glacier-dependent regions is necessary [28].

3.10.4 Policy Integration

International climate assessments (IPCC, the Paris agreement), regional water-management frameworks, and disaster-risk reduction plans all depend on accurate and timely estimates of glacier retreat. One of the fundamental challenges is still communicating irreversible loss and residual uncertainty in a way that facilitates risk-taking decision-making.

3.11 Conclusion

Globally, glaciers are retreating quickly, and due to cumulative warming and climate system inertia, many ice bodies will eventually vanish. Our capacity to predict where and when glaciers may disappear has significantly increased over the last 20 years thanks to developments in satellite surveillance, computational glacier models, and ML. Numerous smaller glaciers may vanish sooner than expected, as observations show that glacier loss is following the high end of previous IPCC estimates.

Predicting glacier loss accurately is not just an academic endeavor; it is a necessary condition for well-informed adaptation in the water, energy, agricultural, and disaster-risk sectors. It is the duty of the glaciological community to provide the best forecasts, presented with clarity regarding both confidence and residual uncertainty, as populations that depend on glaciers face the possibility of disappearing water sources and unstable mountain risks. The next 10 years will be crucial for reducing the socioeconomic effects of the irreversible loss of mountain ice through improved observations, ongoing model development, and close collaboration with decision-makers.

References

[1] IPCC. (2019). Special report on the ocean and cryosphere in a changing climate: Chapter 2 – High mountain areas. Intergovernmental Panel on Climate Change. https://www.ipcc.ch/srocc/chapter/chapter-2/

[2] IPCC. (2021). Climate change 2021: The physical science basis (Working Group I contribution to the Sixth Assessment Report), Chapter 9: Ocean, cryosphere and sea level change. Intergovernmental Panel on Climate Change. https://www.ipcc.ch/report/ar6/wg1/chapter/chapter-9/

[3] Immerzeel, W. W., Lutz, A. F., Andrade, M., Bahl, A., Biemans, H., Bolch, T., ... & Kaser, G. (2020). Importance and vulnerability of the world's water towers. *Nature*, *577*(7790), 364–369. https://doi.org/10.1038/s41586-019-1822-y

[4] Worni, R., Huggel, C., Clague, J. J., Schaub, Y., & Stoffel, M. (2014). Permits and limitations of extensive terrain mapping for glacier hazard and risk assessment: Lessons learned from the 2002 disaster in the Kolka–Karmadon region, Russian Caucasus. *Natural Hazards*, *69*(3), 1547–1570. https://doi.org/10.1007/s11069-013-0770-8

[5] Cuffey, K. M., & Paterson, W. S. B. (2010). *The Physics of Glaciers* (4th ed.). Butterworth-Heinemann.

[6] Cogley, J. G. (2009). Geodetic and direct mass-balance measurements: Comparison and joint analysis. *Annals of Glaciology*, *50*(50), 96–100. https://doi.org/10.3189/172756409787769744

[7] Thibert, E., Blanc, R., Vincent, C., & Eckert, N. (2008). Glaciological and volumetric mass-balance measurements: Error analysis over 51 years for Glacier de Sarennes, French Alps. *Journal of Glaciology*, *54*(188), 522–532. https://doi.org/10.3189/002214308786570296

[8] Pellikka, P. K. E., & Rees, W. G. (Eds.). (2010). *Remote Sensing of Glaciers: Techniques for Topographic, Spatial and Thematic Mapping of Glaciers*. CRC Press.

[9] Rignot, E., Mouginot, J., & Scheuchl, B. (2011). Antarctic grounding line mapping from differential satellite radar interferometry. *Geophysical Research Letters*, *38*, L10504. https://doi.org/10.1029/2011GL047109

[10] Zemp, M., Haeberli, W., Hoelzle, M., & Paul, F. (2006). Alpine glaciers to disappear within decades? *Geophysical Research Letters*, *33*, L13504. https://doi.org/10.1029/2006GL026319

[11] Pfeffer, W. T., Arendt, A. A., Bliss, A., Bolch, T., Cogley, J. G., Gardner, A. S., ... & Moholdt, G. (2014). The Randolph Glacier Inventory: A globally complete inventory of glaciers. *Journal of Glaciology*, *60*(221), 537–552. https://doi.org/10.3189/2014JoG13J176

[12] Hugonnet, R., McNabb, R., Berthier, E., Menounos, B., Nuth, C., Girod, L., ... & Kääb, A. (2021). Accelerated global glacier mass loss in the early twenty-first century. *Nature*, *592*(7856), 726–731. https://doi.org/10.1038/s41586-021-03436-z

[13] Marzeion, B., Cogley, J. G., Richter, K., & Parkes, D. (2014). Attribution of global glacier mass loss to anthropogenic and natural causes. *Science*, *345*(6199), 919–921. https://doi.org/10.1126/science.1254702

[14] Hock, R., Rasul, G., Adler, C., Cáceres, B., Gruber, S., Hirabayashi, Y., ... & Zhang, S. (2019). High mountain areas. In *Climate Change and Land: An IPCC Special Report*. IPCC. https://www.ipcc.ch/srccl/chapter/chapter-2/

[15] Maussion, F., Butenko, A., Champollion, N., Dusch, M., Eis, J., Fourteau, K., ... & Marzeion, B. (2019). The Open Global Glacier Model (OGGM) v1.1. *Geoscientific Model Development*, *12*(3), 909–947. https://doi.org/10.5194/gmd-12-909-2019

[16] Jouvet, G., & Huss, M. (2019). Numerical simulation of Rhonegletscher from 1874 to 2099 using the open-source glacier evolution model OGGM. *Journal of Glaciology*, *65*(251), 401–410. https://doi.org/10.1017/jog.2019.22

[17] Rounce, D. R., Hock, R., & Shean, D. E. (2020). Glacier mass change in High Mountain Asia through 2100 using the open-source Python Glacier Evolution Model. *Frontiers in Earth Science*, *7*, 331. https://doi.org/10.3389/feart.2019.00331

[18] Hands, D. J. (2009). Assessing the generalizability of learned associations. *Journal of Applied Statistics*, *36*(4), 367–387. https://doi.org/10.1080/02664760802554100

[19] Krüger, J., & Scherer, D. (2019). Automatic detection of small-scale snow avalanches using a single Doppler radar. *The Cryosphere*, *13*, 2305–2319. https://doi.org/10.5194/tc-13-2305-2019

[20] Chauhan, S., Scherler, D., Maurer, H., & Langhamer, L. (2025). Machine learning improves seasonal mass balance prediction of glaciers. *The Cryosphere*, *19*, 5801–5820. https://doi.org/10.5194/tc-19-5801-2025

[21] Raissi, M., Perdikaris, P., & Karniadakis, G. E. (2019). Physics-informed neural networks: A deep learning framework for solving forward and inverse problems involving nonlinear partial differential equations. *Journal of Computational Physics*, *378*, 686–707. https://doi.org/10.1016/j.jcp.2018.10.045

[22] Busseau, B. C., Acheson, E., & Jacobs, J. D. (2016). Snowline mapping and remote sensing of glacier extent in the Seward Peninsula, Alaska, 2000–2012. *Remote Sensing Letters*, *7*(1), 38–47. https://doi.org/10.1080/2150704X.2015.1088673

[23] Chen, Y., Deng, S., Wang, Q., Chang, X., & Pan, M. (2025). DL4GAM: A multi-modal deep learning-based framework for glacier accumulation and ablation mapping using multi-source satellite data. *Remote Sensing of Environment*, *312*, 113950. https://doi.org/10.1016/j.rse.2025.113950

[24] Lüthi, S., Grebel, A., & Stocker, T. F. (2024). Deep neural networks for ocean-induced ice-shelf melt rate estimation. *The Cryosphere*, *17*, 499–515. https://doi.org/10.5194/tc-17-499-2024

[25] Rounce, D. R., Hock, R., Maussion, F., Hugonnet, R., Shangguan, D., Qiao, G., ... & Shea, J. M. (2023). Global glacier change in the 21st century: Every increase in temperature matters. *Science*, *379*(6627), 78–83. https://doi.org/10.1126/science.abo1324

[26] NASA Earth Science Division. (2023). *Half of glaciers vanish with 1.5 degrees of warming*. NASA. https://www.nasa.gov/science-research/earth-science/nasa-funded-study-half-of-glaciers-vanish-with-1-5-degrees-of-warming/

[27] Marzeion, B., Kaser, G., Maussion, F., & Champollion, N. (2018). Limited influence of climate change mitigation on short-term glacier mass loss. *Nature Climate Change*, *8*(4), 305–308. https://doi.org/10.1038/s41558-018-0093-1

[28] Nie, Y., Sheng, Y., Liu, Q., Liu, S., Zhang, Y., & Song, C. (2018). A regional-scale assessment of Himalayan glacial lake changes using satellite imagery from 1990 to 2015. *Remote Sensing of Environment*, *189*, 375–396. https://doi.org/10.1016/j.rse.2016.11.008

[29] Fischer, A., Seiser, B., Stocker Waldhuber, M., Mitterer, C., & Abermann, J. (2015). Tracing glacier changes in Austria from the Little Ice Age to the present using a new glacier inventory. *The Cryosphere*, *9*, 753–766. https://doi.org/10.5194/tc-9-753-2015

[30] Rabatel, A., Soruco, A., Francou, B., Richter, M., Sicart, J. E., Favier, V., ... & Wagnon, P. (2013). Current state of glaciers in the tropical Andes: A multi-century perspective on glacier evolution and climate change. *The Cryosphere*, *7*, 81–102. https://doi.org/10.5194/tc-7-81-2013

[31] Bolch, T., Peters, J., Yegorov, A., Pradhan, B., Buchroithner, M., & Blagoveshchenskiy, V. (2011). Identification of potentially dangerous glacial lakes in the northern Tian Shan. *Natural Hazards*, *59*(3), 1691–1714. https://doi.org/10.1007/s11069-011-9860-2

[32] Huss, M., & Hock, R. (2018). Global-scale hydrological response to future glacier mass loss. *Journal of Geophysical Research: Atmospheres*, *123*(1), 323–335. https://doi.org/10.1002/2017JD027000

[33] Ehret, U., Gupta, H., Sivapalan, M., Weijs, S. V., Schymanski, S. J., Blöschl, G., ... & Gong, W. (2014). Advancing catchment hydrology to deal with predictions under change. *Hydrology and Earth System Sciences*, *18*(2), 649–671. https://doi.org/10.5194/hess-18-649-2014

[34] Huss, M., Dhulst, L., & Bauder, A. (2015). New long-term mass-balance series for the Swiss Alps. *Journal of Glaciology*, *61*(227), 551–562. https://doi.org/10.3189/2015JoG15J015

[35] Marzeion, B., Huss, M., Rounce, D. R., & Kaser, G. (2025). Glacier preservation doubled by limiting warming to 1.5 °C. *Science*, *388*(5652), Article eadu4675. https://doi.org/10.1126/science.adu4675

[36] Integrating machine learning with process-based glacio-hydrological models: A hybrid approach for improved glacier runoff projection in northern Sweden. (2025). *Advances in Water Resources*, Article 104xxx. https://doi.org/10.1016/j.advwatres.2025.104xxx

[37] Taloor, A. K., et al. (2025). Sentinel-1A data-based glacier facies identification and mass balance retrieval of Northwestern Himalayan benchmark glaciers: An eight-year analysis (2016–2023). *Progress in Disaster Science*, *25*, 100667.https://doi.org/10.1016/j.pdisas.2025.100667

Ajay Sudhir Bale*, Deepak N. R., Prathima M. G., Chandramouli H., Naveen Ghorpade, Mamta B. Savadatti*, Varun Yogi Shantakumar, Omkar K. Kanetkar

Chapter 4
AI-Powered Gig Work: A New Era of Predictive Analysis

Abstract: This chapter discusses the evolving domain of the gig economy, focusing on how artificial intelligence (AI) is integrated into it and how its effects on both gig workers and businesses. It focuses on the current trends, where the AI and other computational intelligence techniques that reshape work dynamics. This work also focuses on the benefits and related challenges that are faced in the gig economy and how they can be solved using AI, such as increasing the efficiency and flexibility while simultaneously reducing the stress and job insecurity. With the use of case studies, we aim to bring out current trends and the various applications of AI in specific sectors like healthcare, manufacturing and finance. These cases are based on how AI powered platforms are transforming the traditional work systems. The latter part of the chapter also concentrates on the socio-economic implications of these changes, and urges the importance of needs for various regulatory changes and ethical considerations to make sure the gig workers have equitable outcomes. Finally, the work outlines the need for future work and research for this rapid growing field, shedding light on balancing technological innovation and various human-centric models.

Keywords: Algorithmic management, gig economy, job crafting, predictive analysis, human-centric AI

*__Corresponding authors: Ajay Sudhir Bale__, Department of ECE, New Horizon College of Engineering, Bengaluru 560103, Karnataka, India, e-mail: ajaysudhirbale@gmail.com
*__Corresponding authors: Mamta B. Savadatti__, Department of ECE, New Horizon College of Engineering, Bengaluru 560103, Karnataka, India, e-mail: mamta.savadatti@gmail.com
Deepak N. R., Department of Information Science and Engineering, Atria Institute of Technology, Bengaluru, Karnataka, India
Prathima M. G., Department of CSE, Bangalore Institute of Technology, Bengaluru, Karnataka, India
Chandramouli H., Department of CSE, East Point College of Engineering and Technology, Bengaluru, Karnataka, India
Naveen Ghorpade, Department of CSE, Vijay Vittala Institute of Technology, Bengaluru, Karnataka, India
Varun Yogi Shantakumar, Department of Electrical and Electronic Engineering, California State University, Sacramento, CA, USA
Omkar K. Kanetkar, Department of ECE, New Horizon College of Engineering, Bengaluru 560103, Karnataka, India

 | https://doi.org/10.1515/9783112211991-004

4.1 Introduction

A gig economy is a type of open market in which enterprises frequently employ autonomous freelancers for short-term projects, and interim employment is prevalent. A work that endures for a certain duration is sometimes termed a "gig" in informal vernacular. Some examples of gig workers include freelancers, self-employed individuals, project-oriented workers, and transitory or part-time professionals [1, 2]. Customers and gig workers frequently communicate via internet connections and gig apps. The freelance economy is a contemporary phenomenon that has emerged due to many factors. The remaining two elements are of fundamental importance: employment has become more flexible and geographically distributed, with a growing amount of work conducted remotely through digital platforms [3]. Consequently, employment and geographic locations are increasingly becoming disconnected. This indicates that unrelated contractors might accept positions or projects from employers worldwide. Employers may choose the most suitable applicant for a specific job from a broader pool than what is available in a single area. The cooperative economy, shared economy practices, barter systems, and flexible work contribute to the evolving cultural and business environment. The cultural ramifications of the gig economy continue to develop; for example, hiring habits have been profoundly influenced by the COVID-19 pandemic [4]. The flexibility of the gig economy enables individuals to engage in specific activities or tasks offered by companies without requiring a conventional employment framework or compensation packages. The World Economic Forum defines the "gig economy" as a labor market characterized by employees contributing and their capacity to generate income via "gigs" or "tasks" into which they have been contracted [4–8]. This work in [2] determines a gig economy as a type of freelance employment system that connects individuals who leverage their resources, expertise, or abilities to engage in short-term freelance assignments according to particular requirements, with without fixed employment status. This study also illustrates the gig economy environment in Figure 4.1.

The unique employment paradigm of the gig economy, who appears to provide many benefits compared to traditional work, is rapidly transforming the future of labor [9, 10]. Gig workers have more autonomy in selecting employment opportunities as well as scheduling hours, because they are not bound to a specific company and may accept positions from any willing entity. Similarly, organizations and employers gain from greater flexibility, as they become less dependent on a static crew to execute tasks. They may also economize since they are not obligated to give workers the same benefits, advancements, protection, etc., that are customary in regular employment. The "gig economy" has lately proliferated globally. Approximately 25% of all US workers engage in gig labor, with 10% depending only on gigs for their income. However, the gig economy has disadvantages too. Gig workers possess more flexibility regarding work and hours, enhanced freedom in executing tasks, a broader array of gig opportunities, and the possibility of increased profits via incen-

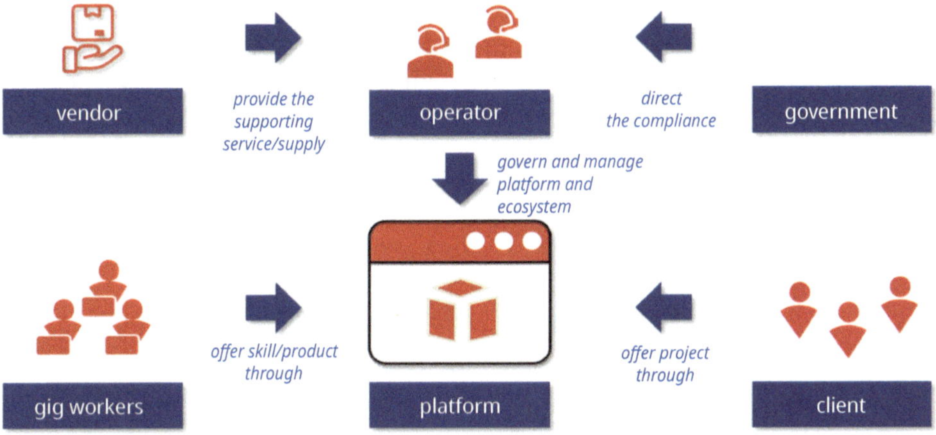

Figure 4.1: Ecosystems of the gig sector [2]. Reproduced from open access journal.

tives. Nevertheless, they encounter challenges, including heightened levels of tension and diminished monetary security due to being unable to eliminate work-related expenses from their salary [11, 12]. The benefits of the gig economy for enterprises are evident – reduced overhead expenses (pay per task), swift capacity, and entry to a variety of open labor. Yet it also presents disadvantages, such as unreliable personnel and legal and regulatory concerns regarding worker classification. The demand-and-supply metrics can alter these advantages and disadvantages for corporations and workers. To enhance their revenue, many individuals in India – where the unemployment rate is very high – use carpooling firms such as Uber and Ola [13–15]. Despite the increase in the workforce, companies often reduce compensation while increasing working hours. The health of workers may deteriorate due to increased stress, and the frequency of accident risks may increase due to worker fatigue and stress [16]. Furthermore, as laborers are not classified as workers, the companies engaged are not able to be held liable for any harm that may arise [1]. This circumstance pertains strictly to low-skilled employment, that was prevalent notwithstanding market limitations. Professionals' positions and contracts are highly sought after but scarce, resulting in elevated compensation for entrepreneurs in these domains. Consequently, employers possess more authority in selecting candidates for inexperienced positions, while they have greater power in specialized consulting sectors [17, 18]. The use of freelance employment within the medical field is a compelling topic for further investigation. For example, Saudi Arabia's healthcare system is rapidly evolving due to Vision 2030. Nevertheless, the nation's healthcare sector is mostly reliant on foreign labor. The predominant segment of healthcare personnel in Saudi Arabia comprises nurses, who are crucial to overseeing the everyday functions of several healthcare facilities and serve as the foundation of the healthcare system [19, 20]. In 2018, Saudi Arabia's ratio of 5.5 nurses per 1,000 people surpassed the average among other MENA nations. Nonetheless, the recruitment of foreign nurses has risen in the past few years. The nursing profession exhibits a gender imbalance issues, female nurses attending to male patients may not have familial support, many nurses get inadequate compensation and endure extended working hours, and opportunities for advancement are limited. The next two sections discuss about the recent trends that have provided solutions for the aforementioned problems. Some cases are also discussed to show how they are used to solve these problems.

4.2 Current Trends

In light of current advancements in artificial neural networks (ANNs) and their extensive capacities, the study in [1] aims to use these networks to predict workforce alignment with the gig economy framework (GEF). They collected and disseminated exceptional data. This study attempted to create a neural network for predicting interviewees' compliance with the GEF by analyzing their replies to demographic and economic questionnaires. The data are input into a multilayer perceptron (MLP) neural network in the subsequent part. The overarching tendency of the present paper in [1] is illustrated in Figure 4.2.

The ANN included 20 inputs, including characteristics such as age, gender, level of education, occupation, previous employment, and ethnic background, with answers to 14 survey enquiries. The system's output addressed the query about the impact of the gig economy on four areas of human life: personal life, personal growth, career advancement, and managing money. This study used 379 samples, including 267 to train the network, 57 for validation, and 57 for the final test. The accessible data consisted of qualitative content that was transformed into numbers for utilization. The produced data comprises five distinct classes. Figure 4.3 illustrates the arrangement of the previously discussed network.

4.2.1 Hypothesis Model

Gig workers use computational software for management to get relevant assets, like labor service-specific assignments, and client assessments, to elucidate which behaviors are acknowledged by the website. Consequently, people react favorably to the technological assistance offered by the algorithm system, enhancing their job effectiveness by acquiring new abilities and amassing resources for career-oriented job design. Furthermore, automated systems continuously update achievement rankings on the work page and offer operational promotions, fostering a workplace where commendable efforts are honored by the tool and acknowledged positively by peers,

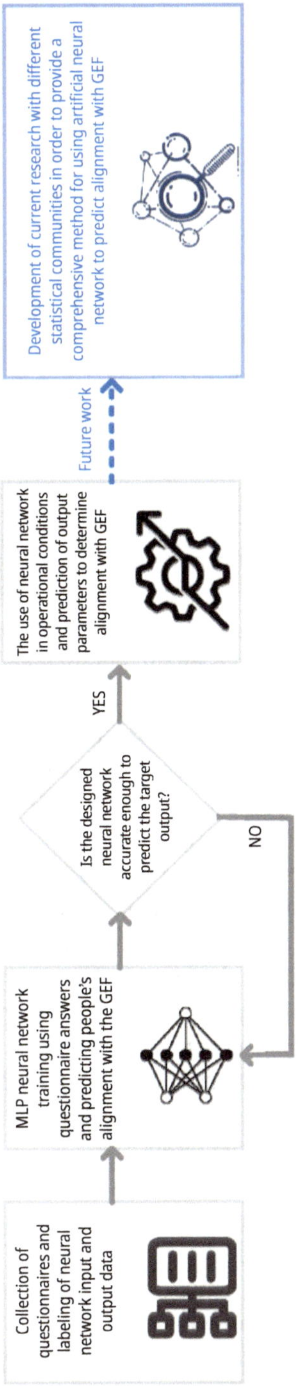

Figure 4.2: Workflow in [1]. Reproduced from open access article.

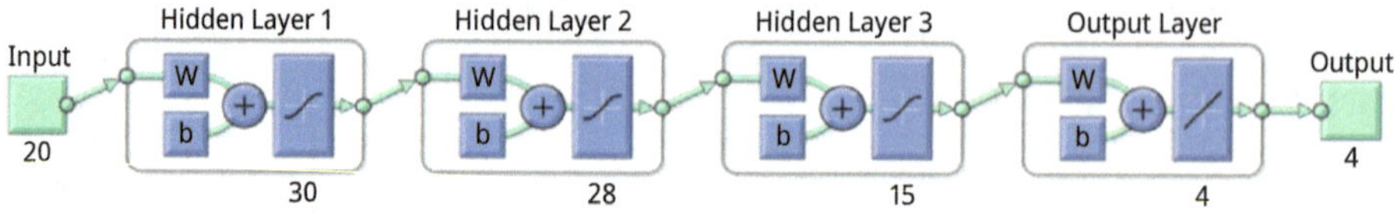

Figure 4.3: MLP in [1]. Reproduced from open access article.

thereby enhancing the engagement of freelance workers, reinforcing career drive, and encouraging career advancement behaviors. Consequently, a particular hypothesis is posited in [11].

4.2.2 Job Crafting

Concurrently, algorithmic technologies offer ongoing monitoring and action via algorithmic controls and standardized procedures integrated into the platform [1]. The messages of information sent by these management strategies engender sensations of scarcity and uncertainty about their jobs among gig workers, consequently adversely affecting their behavioral reactions and prompting risk-avoidance job crafting. Secondly, the methods effectively integrate multifaceted variables to autonomously ascertain the job evaluations and compensation of gig workers [11]. This form of transparent decision-making frequently causes gig workers to perceive automated leadership as disrespectful, often resulting in adverse attitudes and an uptick in prevention-focused career development, such as deliberately filtering clients and disregarding algorithmic prompts [21–23]. The job demands-resources (JD-R) approach classifies working characteristics into two primary classifications: job demands and job resources. Job demands pertain to the individual's physical as well as mental capabilities required at work, including workloads, time constraints, dealing with difficult clients, inconvenient timetables, and adverse physical conditions. This hypothesis asserts that extremely high job expectations could end up in burnout, leading to detrimental work behaviors. Under the oversight of the computational structure, the labor processes of gig workers are entirely transparent, compelling them to conform to computational standards and achieve unrealistic work objectives through excessive workloads, thereby entering a state of diminishing resources.

The study in [11] enhances the job crafting study's structure and expands the definition of job crafting to include this technology, informed by the dual-action approach of algorithmic management. This is a novel examination of the influence of algorithmic leadership on the job-building behaviors of gig workers, hence enhancing the foundational study on job crafting within the gig economy. Moreover, the majority of pertinent research has not thoroughly investigated the facets of work making, mostly concentrating on promotion-focused job crafting while neglecting prevention-focused work crafting [24, 25]. This study combines the two aspects of job crafting and examines the relevant antecedent factors individually. Moreover, prior studies have mostly investigated the processes of job crafting using several conceptual structures, including the job demands–resources framework, sustainability resource theory, and theories of self-determination. This study employs the principles of social computing theory to examine job crafting through the stages of content intake, interpreting, and output, therefore enhancing the theoretical framework of job crafting investigation. This work broadens the research viewpoint regarding the moderating processes of algorithmic control efficacy. The overall study is depicted in Figure 4.4.

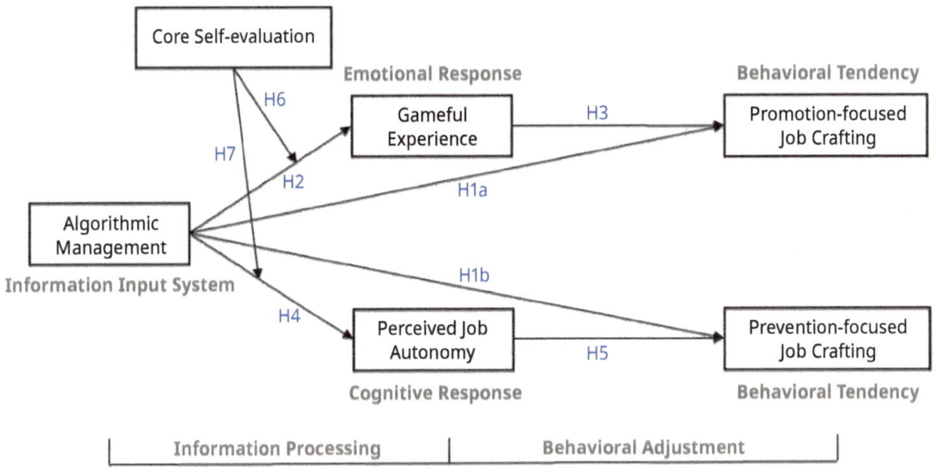

Figure 4.4: Model in [11]. Reproduced from open access journal.

The scholarly investigation into the mechanics of influence in computerized leadership is still in its nascent stages. While existing research has in theory suggested the effectiveness of the "hybridity" viewpoint on computational management, the precise mechanisms by which algorithmic leadership rapidly and intricately affect the work results for gig workers necessitate further investigation within particular leadership situations. Social information processing theory offers an appropriate mathematical structure for this study, enabling an analysis of the dual impacts of computational management. The authors examined the "black box" coupling of algorithmic management and promotion-focused career-building practices from psychological and behavioral perspectives, therefore enhancing empirical studies on the impact mechanisms of algorithm-driven management systems. Ultimately, basic self-assessment is a crucial prerequisite for algorithmic administration.

While prior studies often acknowledge the significance of workers' core self-evaluation about their job capability and drive, they rarely address it within a particular managerial framework. This research investigates and confirms the limiting impact of a fundamental self-assessment among freelance employees, elucidating its effect on the interaction between computational leadership and job design. The results address the critical inquiry about the timing for promoting the beneficial aspects of algorithmic management while mitigating its adverse impacts. This study provides new perspectives on individual handling of algorithmic management effects and broadens the research area on fundamental self-evaluation in organizational leadership [26].

4.2.3 Algorithmic Pricing and Time-Related Stress

The gig economy provides its staff members with significant schedule flexibility, which appears to be the most appealing aspect of gig employment, resulting in heightened happiness at work. This presents a contradiction, since evidence indicates that algorithmic management diminishes worker autonomy owing to externally established objectives, stringent time limitations, and heightened surveillance. The properties of algorithmic compensation seem to amplify time constraints and work magnitude, which are linked to time-related stress. Time-based stress denotes concern stemming from workers' perceptions of heightened job responsibilities due to insufficient time to perform tasks. Moreover, several scholars assert that gig labor resembles piece-rate employment, a rigorous variant of pay for performance. Piece-rate compensation permits online labor platforms to remunerate solely for "efficient" time, causing gig workers to endure prolonged periods without pay while seeking employment, which leads to extended hours of labor and encourages the acceptance of numerous gigs, including those that are undesirable or hazardous [27].

Meal delivery mechanisms provide gig workers with a flexible schedule; yet, they continuously assess, direct, and restrict visits to standardize service, thereby penalizing workers for failing to reach performance standards. Moreover, these regulatory systems may malfunction, directing cyclists into automobile-exclusive lanes or omitting GPS information, which hinders workers from obtaining their deliveries, consequently extending travel times and compromising client satisfaction and efficiency objectives. Although gig workers possess minimal authority concerning these algorithmic remuneration challenges, they may still face repercussions as a result. Research indicates that delays in job completion may be attributed to restaurants failing to provide orders punctually, gig workers experiencing theft of their transportation, consumer behaviors, and the inherent processes of the platforms themselves. Moreover, the algorithmically assigned targets must be accomplished under inclement weather, hence increasing the danger of mishaps. Notwithstanding these concerns, few data collection seems to have assessed the effects of computational reimbursement on temporal stress and employee health [12].

4.3 Case Study

The following cases developed a hybrid gig work platform that combines artificial intelligence (AI) technology with cross-domain expertise.

4.3.1 Hybrid Gig Work Platform for Medical Professionals

There is unanimous agreement about the extraordinary potential of AI. In the years to come, it will transform every component of our lives, including healthcare. Nevertheless, several individuals have apprehensions over AI's potential dominance over humanity. Stephen Hawking stated that the development of comprehensive AI may signify the demise of the human species. However, if mankind is sufficiently ready for the new age of AI, it will emerge as the next effective domain of collaboration between people and machines. AI will fundamentally transform healthcare for the greater good. AI may assist healthcare practitioners in formulating treatment strategies and identifying the most appropriate methodologies for each patient. It may facilitate repetitive, tiresome tasks, allowing doctors and nurses to focus on their primary responsibilities rather than contending with bureaucratic processes [21, 22]. In the era of Big Data, the value of medical data is indisputable [22]. The involvement of technology giants like Google or IBM in the clinical analysis of data underlines its substantial importance [28]. In 2019, DeepMind's healthcare division amalgamated with Google Health to "develop solutions that assist healthcare providers and enhance patient results." Google Health is using AI to enhance the detection of lung and breast cancer screenings, forecast outcomes for patients, prevent blindness, and more. These are not just platitudes; Google has shown its commitment via action. In collaboration with its DeepMind division, Google Health has developed an AI-driven technology for the detection of breast cancer. Furthermore, the software surpassed all of the human radiologists it was compared with, annually by 11.5%!

Clinical Information Expert Augmedix created methods for gathering information from authentic physician-patient dialogues and transforming it into clinical notes instantaneously, which are then sent to the electronic health record systems of medical professionals. Mitigating the managerial duties faced by healthcare staff and automating these tasks can significantly save time in the industry and may also alleviate signs of burnout among healthcare workers. Their recent news was the introduction of a completely automatic, generative AI (GenAI)-driven medical record solution for Emergency Departments [29, 30]. This enhances the Augmedix Go solution, a clinician-operated mobile application that uses GenAI to promptly generate a automatically generated draft clinical report after every individual consultation. The firm released survey findings indicating that practitioners in ambulatory care environments may save one hour or more every clinic day by using their technology [23].

Their AI-driven platform employs ambient speech technologies and speech recognition to streamline the generation of healthcare notes. DeepScribe employs sophisticated speech recognition algorithms to transcribe real dialogues between medical professionals and patients into detailed, precise medical records. The EHR connectivity enables doctors to promptly evaluate and approve the notes. The business claims that its technology conserves up to 3 h of time for physicians daily. DeepScribe claims to have developed its program using the biggest library of authentic patients' dialogues globally, claiming that it is 32% higher in precision than GPT-4. It is important to acknowledge that GPT was not explicitly trained for use in medicine [24].

4.3.2 Gig Work in Improving the Quality Control in Workplaces Using AI

The incorporation of AI tools and equipment in businesses presents substantial opportunities for advancement and efficiency enhancement. Nonetheless, significant occupational safety and health (OSH) concerns emerge when AI is incorporated into work environments. Stress, bias, increased precariousness, musculoskeletal disorders, and the potential for work intensity and layoffs have been shown to present psychological dangers, involving physical assault in digitized workplaces. These concerns are intensified when AI enhances pre-existing technological instruments or is newly used for managerial tasks and design. AI indeed amplifies occupational safety and health concerns in digitalized workplaces by facilitating better oversight and tracking, perhaps resulting in overreach, a primary contributor to stress and anxiety. AI emphasizes the need for enhancing confidence and possible authority of predictive AI systems, robots, and algorithmic methods in the workplace. However, it is important to emphasize that technology alone does not provide occupational safety and health advantages or dangers. The application of technology generates either negative or favorable situations [25].

4.3.3 In Manufacturing

AI is significantly transforming manufacturing, allowing enterprises to attain superior quality requirements, greater effectiveness, and more innovative use of resources [26]. Each manufacturer is well aware of the challenges posed by unanticipated downtime: the complete cessation of manufacturing, the urgent effort to diagnose the issue, and the hours or days wasted. Envision a device capable of forecasting machinery malfunctions prior to their occurrence, enabling proactive intervention. AI-driven maintenance planning is effectively doing this, saving millions for firms and ensuring uninterrupted manufacturing lines. The next phase of industrialization transcends mere speed; it emphasizes intelligence in production. The most recent Deloitte survey [27] revealed that 86% of company managers anticipate that advanced technological advancements, incorporating AI, will enhance competitiveness over the subsequent 5 years. Factories are using modern technologies like predictive maintenance and real-time monitoring of quality to save costs, minimize interruptions, and attain elevated levels of accuracy and dependability. AI is transforming the methods by which manufacturers address the difficulties of contemporary marketplaces. Six methods by which AI improves industrial processes are discussed below.

4.3.3.1 Predictive Maintenance with AI-Driven Analytics

These solutions are transforming equipment management for factories, minimizing downtime, lowering expenses, and maintaining uniform product quality. Through the analysis of sensor data – including temperatures, noise, and movement – alongside outside forces such as weather, AI forecasts the need for equipment upkeep prior to the emergence of problems. This preventive approach enhances equipment longevity, optimizes assets, and elevates the level of the product. Properly kept technology functions at optimal efficiency, yielding regular and superior results. A commercial blender optimized by AI-generated insights produces consistent blends, yielding enhanced final goods. In addition to superior performance and effectiveness, AI also improves worker safety. Predictive and preventive methods to avoid machinery breakdowns drastically decrease the likelihood of mishaps resulting from unforeseen malfunctions, hence fostering a safer workplace for employees.

4.3.3.2 Real-Time Quality Assurance

Conventional methods for quality control often include mechanical checks that can be labor-intensive, susceptible to inaccuracies, and may overlook significant faults. AI-driven identification of anomalies provides a contemporary solution, detecting abnormalities and discrepancies in real time to ensure goods adhere to rigorous quality requirements. In 2024, 63% of production firms indicated the use of AI for quality assurance, with applicability extending to maintenance planning and real-time optimization of processes. A comprehensive quality management suite (QMS) significantly augments this methodology by offering effective tools for overseeing and monitoring quality across manufacturing processes. Quality management system solutions enable firms to enhance defect identification with ML algorithms that spot discrepancies promptly. QMS solutions facilitate evaluations of quality, allowing producers to continue in standards with less effort [31].

4.3.3.3 Managing of Inventory

AI algorithms facilitate realistic forecasting of demand via the analysis of past data, developments in the market, and consumer behavior to predict future requirements. Comprehending future requirements enables firms to optimize inventory levels and mitigate overstock and stockouts. AI-driven automatic restocking solutions enhance productivity by optimizing order fulfillment and minimizing human involvement. AI empowers firms to optimize year-round activities, reduce waste, and enhance revenue via trend-based forecasting systems that account for seasonal and market changes [32].

4.3.3.4 Autonomous Manufacturing

AI-driven automation decreases the involvement of humans in production processes, hence improving efficiency and uniformity. The use of robots enables organizations to perpetually execute operations such as installation, welding, and packing using AI-driven networks, guaranteeing consistent quality and enhancing efficiency. Collaborative robots, or cobots, operate in conjunction with people to enhance both productivity and security by handling components, conducting quality checks, and executing various extremely monotonous jobs. These advancements enable companies to optimize processes while ensuring accuracy and dependability [33].

4.3.3.5 Optimization of Labor Management

Labor expenses constitute a substantial cost for producers; however, AI is facilitating the optimization of labor administration. Through the analysis of production requirements, worker accessibility, and past achievement metrics, AI generates optimized timetables designed to fulfill corporate objectives. Forecasting workforce requirements guarantees optimal resource allocation at appropriate times, hence enhancing efficiency on the work floor. AI detects skill deficiencies, allowing focused training initiatives that enhance staff competencies, including compliance and superior training. Furthermore, AI decreases injury rates and fosters better safety at work by automating high-risk jobs, hence enhancing worker satisfaction and reducing turnover [34].

4.3.3.6 Expense Minimization and Productivity Enhancements

AI decreases dependence on human labor by automating mundane tasks and improving the distribution of resources, hence lowering operating expenses. Companies gain from optimized processes, reduced waste, and enhanced production, all of which bolster financial performance. Businesses are rethinking operational possibilities using AI [35].

4.3.4 Gig Platform for Economic Analysis Driven by AI in the Finance Industry

Millions of undocumented immigrants are signing up for gig labor platforms in developing economies to do things like deliver packages or provide household, specific, or business services. As their conditions of employment come under much-needed examination, gig networks are poised to become a major sources of income for those with limited resources throughout the world. Even if they are entering digitally formalized labor, the majority of workers in developing economies that join platforms are still severely underpaid. They post job openings, get payments into online accounts, and generate large volumes of data on the frequency, excellence, and amount of their labor on sites. Their access to financial services is expanded via new avenues made possible by the digital link and the data produced [28]. Fintechs and FSPs have direct exposure to big groups of hard-to-reach clients with digital financial institutions and verifiable earnings data when they collaborate with gig marketplaces. Financial support for employees may help platforms boost retention and participation rates by lowering churn and making the website's environment more stable and efficient. Services and fintechs have begun to provide independent contractors in many developing countries with a range of savings, insurance, and credit services via their platforms because they saw a possibility. In order to establish an educational environment centered on this cutting-edge area of innovation, CGAP collaborated with fintechs and gig platforms across the last 2 years to evaluate the efficacy and worth of these kinds of financial services for freelancers. Through assisting with funding access, emergency management, and revenue smoothing, the evidence suggests that these kinds of services may allow gig workers' financial autonomy while helping their daily lives. The full potential of financial platforms is still limited by a number of issues. In order to do anything about these obstacles and the current level of innovation in the field, financiers and the wider crowdfunding industry need to know what's happening. To increase the likelihood that this environment will be beneficial to employees, we provide some observations and recommendations for next steps [36].

During a recent CGAP gathering of 50 leaders from platforms and fintechs, there was a pronounced consensus that creativity is progressing well; nonetheless, mechanisms to transform are essential to fully realize the system's potential as an intermediary. Similar to previous instances of emerging innovation, such as mobile banking or electronic micro-credit, financiers and the monetary sector may facilitate market growth and technological advancement while assisting authorities and regulators. The financial autonomy society's objectives are as follows:

A. Assist infrastructure and fintech companies in testing innovative solutions. Interfaces and fintechs need enhanced expertise in implementing these solutions to demonstrate the correlation among banking and insurance and retained staff. Initial encounters have disheartened several individuals, since exclusively online offerings haven't culminated in substantial adoption. The duration needed to consolidate these business model may occasionally align with the rapid, high-churn nature of platform companies. Investors may facilitate experiment and insight-gathering via patient investment.
B. Encourage the finance sector to adopt innovative models with confidence. Diverse stakeholders in the financial industry, particularly credit insurers, conventional banks, and regulatory bodies, must cultivate confidence in novel credit rating and banking frameworks. Funders may facilitate the dissemination of insights on the effective use of metrics by innovators, hence enhancing public confidence.
C. Assist governing bodies in optimizing the implementation of work data. Governments might need to take steps to enhance the quality and accessibility of employment data, ensuring its use and reliability. Ultimately, it may be beneficial to regard employment data similarly to money data and integrate it with credit reporting agency information. This would enable workers to organize their data in a trustworthy, transparent, and consistent manner, thereby enhancing their ability to connect to innovative and further resilient goods. At present, this knowledge is possessed by various channels, each acquiring a portion of revenue and financial data. Financiers may orchestrate dialogues between governments and digital platforms to enable these modifications.
D. Assist policymakers in establishing accountable relationships with platforms. Authorities aiming to safeguard the interests of these workers may discover that portals offer an effective means of engaging with them, therefore facilitating access to social security systems as well as emergency assistance for this demographic. This creates opportunities for collaborations between governments and networks to enhance assistance for this sector.
E. In the future, it can be expected that digital methods of arranging employment and receiving payment will generate greater volumes of data from diverse sources and enhance communication via the internet. Utilizing that data and those methods to engage marginalized sectors will need novel engagements and collaborations. Insights gained from delivering financial assistance to gig workers may provide valuable insights relevant to other marginalized populations, like agricultural laborers, rural employees, and women, who are progressively engaging in digitally mediated employment. Consequently, it is essential for the financial security sector to collaborate effectively with the commercial sector and govern-

mental entities to enhance the impact of financial inclusion for those working on platforms.

4.3.5 Creative Sectors: AI-Enhanced Design and Content Production

Creativity is fundamental to humanity. GenAI, such as robust large language models (LLMs), has the potential for enhanced human creativity via the provision of novel concepts, or diminished creativity by relying on AI-generated ideas. Large language models such as ChatGPT are increasingly being used in software freelancing, automated coding support, and AI-assisted digital work environments within the gig economy [8]. The work in [29] investigated the causal effect of GenAI concepts on the creation of short tales using an internet-based study as shown in Figure 4.5, in which some authors received narrative ideas from an LLM. Exposure to GenAI concepts leads to tales being seen as better structured, more composed, and more pleasant, particularly among less imaginative authors.

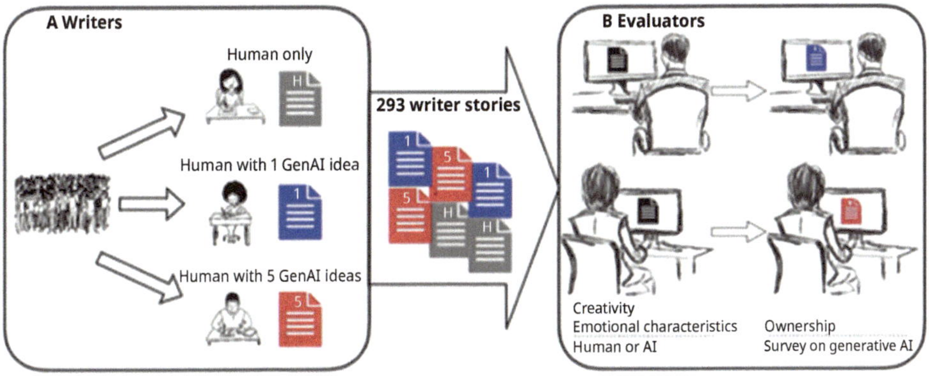

Figure 4.5: Graphical depiction of design [29]. Reproduced from open access journal.

Nonetheless, narratives generated by AI have more similarity to one another than those created just by humans. These findings indicate a rise in personal imagination at the expense of collective creativity. This dynamic mirrors a societal problem: while GenAI enhances solitary authors' outcomes, it results in a substantially diminished range of innovative output. Our findings have significance for scholars, lawmakers, and professionals aiming to enhance innovation [29].

4.3.6 Agriculture: AI-Enhanced Crop Surveillance and Evaluation

Shortages of labor in agriculture constitute a worldwide issue that results in revenue declines and resources inefficiencies. In the United States, migration constitutes the primary supply of laborers; however, labor immigration has fallen by 75% in recent years, while the financial losses from uncultivated crops owing to labor shortages for farming were calculated at USD 3.1 billion annually in 2014. The work in [30] examines the ongoing shortages of labor in agriculture in the southern United States and Mexico, assessing the viability of mitigating the shortages using a labor matchmaking system augmented by AI. The research analyzes the monetary implications and architectural inadequacies in agricultural labor sectors [31–33], investigating whether a digital medium might reconcile the disparity between labor demand and supply in agriculture. The study uses a multifaceted methodology including a comprehensive literature analysis, detailed user surveys, system dynamics modeling, and empirical investigation related to the establishment of a firm and the introduction of a minimum viable product (MVP). The MVP, a fundamental element of the planned online platform, has undergone market testing to collect quantifiable information and insights via online advertising [34–36]. The results underscore the platform's ability to optimize labor matching procedures, enhance openness, and augment efficiency within the agricultural labor market. Furthermore, the use of AI enables sophisticated matching abilities, accurately forecasting and aligning labor requirements with available workers. This work in [30] offers an economically feasible company structure to address a significant economic issue and adds to the wider discussion on the technological role in altering conventional sectors in both industrialized and developing nations [30].

4.4 Challenges and Possible Solutions

Deploying AI solutions sometimes requires significant initial expenditures in hardware, specialized programs, and network enhancements. For medium-sized businesses, these expenses may be burdensome, hindering uptake notwithstanding the potential benefits of AI. Companies may use flexible enterprise management platforms equipped with incorporated AI functionalities to mitigate substantial startup expenses. Cloud-based AI systems expand in conjunction with their business activities, offering a cost-efficient solution that obviates the requirement for in-house equipment.

AI systems excel with substantial amounts of high-quality data; nevertheless, good data management sometimes presents difficulties. Organizations need resilient systems to gather, retain, and manage data to guarantee its accuracy and relevance. Companies may standardize data cleansing, organization, and preservation methods by using information control systems. Those structures assist manufacturers in guaranteeing that data are unified, correct, and readily usable for AI research.

Numerous facilities depend on antiquated equipment that is incompatible with contemporary AI technology. Retrofitting older systems to integrate with AI solutions may be intricate and costly, sometimes necessitating comprehensive testing and alterations. Companies may address this gap by using company administration solutions equipped with integrated APIs and adaptable connection functionalities. Staged integration solutions enable firms to progressively modernize their systems, hence reducing interruptions to current operations.

Educating personnel to use AI technologies, analyze findings from data, and manage automated systems may be laborious and resource-intensive. Manufacturers should engage in ongoing education and growth initiatives to tackle this difficulty. These programs assist workers in developing the technical proficiency required to run AI systems, while also fostering their confidence and empowerment in adjusting to their changing responsibilities.

AI systems, managing significant amounts of private data about operations, inherently draw attackers. The increasing use of generative AI and automated decision systems in digital platforms also raises concerns regarding privacy, ethical data handling, and algorithmic transparency [5]. AI assists firms in addressing these concerns by facilitating strong security measures that protect essential data. Instruments such advanced encryption, multi-tiered access restrictions, and automated audits identify weaknesses and thwart unauthorized access. AI-powered compliance monitoring guarantees conformity with rules like GDPR and CCPA, enhancing data security while minimizing the risk of expensive compromises. By using these AI-driven protections, businesses fortify their infrastructure and establish confidence with customers, guaranteeing their activities remain reliable and lawful.

4.5 Discussions and Future Scope

AI has evolved in the gig economy as it represents the transformational shift that has gone beyond the technological innovation, as shown in this chapter through multiple sector-based studies. AI is found to be useful in reshaping how work is perceived as a concept, how it is managed and finally how it is executed across various industries. This chapter highlights different types of methods and technological potential with human experience and it presents both remarkable opportunities and the challenges that are present. In healthcare, it demonstrates extraordinary potential by which the administrative processes and professional burnout can be elevated. Technologies like Augmedix and DeepScribe have shown how this technology can dramatically reduce physicians' workloads and save them multiple hours daily. These innovations are not only used to improve operational efficiency but are also used by healthcare professionals to redirect their focus to patient care, which will represent a critical advancement in medical history. In the manufacturing sector, another main experience is transformation with 86% of managers anticipating in technology which will further enhance the effectiveness of the system.

The creative sector shows that the literature extensively discusses of AI's impact on human creativity. While these techniques may give creative output but they are less imaginative and several literary works have shown that this simultaneously poses a risk where the collective creativity and diversity will be reduced. So, this dynamic work presents a fascinating way of showing how the societal challenges can be resolved without compromising the unpredictable nature of human creative expression. In the domain of agricultural applications, the problems of global labor challenges can be solved. Further, when these techniques are used in this domain the shortage of labor and optimization of production will become very easy. It further answers and transform the agriculture productivity. These innovations are very much important in places where migration and declines in labor availability are prevalent as they provide the technological bridge that will address these economic and operational inefficiencies.

Robust security measures and ethical considerations are very important because AI systems are increasingly used to handle sensitive data developed using comprehensive protection techniques and ensuring that the compliance with regulations is maintained. This involves creating transparent and fair algorithmic management design systems that respect worker dignity and rights, apart from technological safeguards. The interdisciplinary research also plays a critical role in understanding AI's broad societal implications. Emerging technologies such as blockchain-integrated augmented reality systems may further enhance remote collaboration, workforce training, and immersive management within future gig ecosystems [7]. Further investigations in the future should explore the economic, sociological, and psychological impacts of AI-driven work space technology. All this requires collaborative efforts in social sciences, economical, and policymaking to develop human-centric approaches. Apart from technological replacement, the goal is to create an adaptive and supportive work environment. Immersive technologies such as augmented and virtual reality may further improve workforce training, collaborative learning, and knowledge-sharing in AI-enabled gig ecosystems [9]. Success will depend on striking a delicate balance between human experience and innovation, which further ensures that AI serves as a collaborative tool that enhances the above features rather than diminishing the potential of humans. As this gig economy evolves in the future, the AI stands as a transformative force that reshapes not just how we work but how we conceptualize the new work environment.

References

[1] AL-Dossary, R., Mayet, A. M., Bhutto, J. K., Shukla, N. K., Nazemi, E., & Qaisi, R. M. A. (2023). Proposing a method based on artificial neural network for predicting alignment between the saudi nursing workforce and the gig framework. *Sustainability*, *15*, 12728. https://doi.org/10.3390/su151712728

[2] Gandhi, A., & Sucahyo, Y. G. (2021). Architecting an advanced maturity model for business processes in the gig economy: A platform-based project standardization. *Economies*, *9*, 176. https://doi.org/10.3390/economies9040176

[3] Deshwal, K. (2025). The role of artificial intelligence in the gig economy's transformation. International Journal of Research and Analytical Reviews, *12*(1), 218–227

[4] Ghorpade, N., et al. (2024). "Acute lymphoblastic leukemia detection employing deep learning and transfer learning techniques." 2024 International Conference on Advances in Computing, Communication and Applied Informatics (ACCAI). IEEE.

[5] Bale, A. S., et al. (2024). The impact of generative content on individuals privacy and ethical concerns. *International Journal of Intelligent Systems and Applications in Engineering*, *12*(1s), 697–703.

[6] Alauddin, F. D., et al. (2025). The influence of digital platforms on gig workers: A systematic literature review. Heliyon, 11(2), e40893.

[7] Hashim, M. F., Bundele, K. S., & Vaishnav, J. (2023). 9 applications and future trends of artificial intelligence and blockchain-powered augmented reality. *Handbook of Augmented and Virtual Reality 1*, 137.

[8] Vada, Y. R., Oshiojum, B. E., Lakkineni, U. K., Chandra Rao, K. V., & Rani, I. (2023). ChatGPT in software development: Methods and cross-domain applications. *International Journal of Intelligent Systems and Applications in Engineering*, *11*(9s), 636–643.

[9] Bale, A. S., et al. (2024). "Augmenting knowledge management with immersive technologies: Exploring transformative use cases for AR and VR." 2024 International Conference on Advances in Computing, Communication and Applied Informatics (ACCAI). IEEE.

[10] Lee, E., & Mitson, R. (2025). Understanding gig workers' perceptions of organizational authenticity and transparency through AI management. *International Journal of Strategic Communication*, *19*, 1–31. https://doi.org/10.1080/1553118X.2025.2470880

[11] Liu, R., & Yin, H. (2024). How algorithmic management influences gig workers' job crafting. *Behavioral Sciences*, *14*, 952. https://doi.org/10.3390/bs14100952

[12] Semujanga, B., & Parent-Rocheleau, X. (2024). Time-based stress and procedural justice: Can transparency mitigate the effects of algorithmic compensation in gig work?. *International Journal of Environmental Research and Public Health*, *21*(86). https://doi.org/10.3390/ijerph21010086

[13] Khang, A., et al. eds. (2025). *Synergy of AI and Fintech in the Digital Gig Economy*. CRC Press.

[14] Talukder, M. B., Chowdhury, S. A., & Khan, M. R. (2025). The gig economy: economic innovations and technological advancements in human resource management. In *Economic Innovations and Technological Developments in HRM*. IGI Global, 381–404.

[15] Khang, A. (2025). Planning for the gig economy era. In *Synergy of AI and Fintech in the Digital Gig Economy*, 81.

[16] de Souza Arruda, M. (2025). *Chatbots in Paradigm Shifting for Improving Gig Workers' Labor* (vol. 9). Proceedings of Ninth International Congress on Information and Communication Technology: ICICT 2024. London: Springer Nature.

[17] Sarala, R. M., et al. (2025). Advancing research on the future of work in the age of Artificial Intelligence (AI). *Journal of Management Studies*, *62*(5). https://doi.org/10.1111/joms.13195

[18] Jabagi, N., et al. (2025). Do algorithms play fair? Analysing the perceived fairness of HR-decisions made by algorithms and their impacts on gig-workers. *The International Journal of Human Resource Management*, *36*(2), 235–274.

[19] Sivaprakasam, A., & Jazil, M. (2024). “Revolutionizing the gig workforce: Technological shifts and emerging opportunities.” 2025 6th International Conference on Mobile Computing and Sustainable Informatics (ICMCSI). IEEE.

[20] Sridhar, V., et al. (2025). Modelling and analysis of the market dynamics and regulation of online gig work. *Digital Policy, Regulation and Governance*, *27*(1), 74–93.

[21] Top artificial intelligence companies in healthcare. (2024, August 13). The Medical Futurist. https://medicalfuturist.com/top-artificial-intelligence-companies-in-healthcare/

[22] DeepMind’s health team joins Google Health. (n.d.). Google DeepMind. Retrieved March 26, 2025, from https://deepmind.google/discover/blog/deepminds-health-team-joins-google-health/

[23] Resources. (2024). Augmedix.com. Retrieved March 26, 2025, from https://www.augmedix.com/resources

[24] Medical scribe software gives doctors hours back in their day. (2024). Deepscribe.Ai. Retrieved March 26, 2025, from https://www.deepscribe.ai/resources/medical-scribe-software-gives-doctors-hours-back-in-their-day-heres-how-theyre-spending-it

[25] Moore, P. V. (2019). OSH and the future of work: Benefits and risks of artificial intelligence tools in workplaces In Duffy, V. (eds). *Digital Human Modeling and Applications in Health, Safety, Ergonomics and Risk Management. Human Body and Motion. HCII 2019. Lecture Notes in Computer Science()* (vol. 11581). Springer: Cham. https://doi.org/10.1007/978-3-030-22216-1_22

[26] Baldwin, D. (2025, February 26). AI in manufacturing: Reshaping quality control and efficiency. Quality Magazine. https://www.qualitymag.com/articles/98588-ai-in-manufacturing-reshaping-quality-control-and-efficiency

[27] 2025 manufacturing industry outlook. (2024, November 19). Deloitte Insights; Deloitte. https://www2.deloitte.com/us/en/insights/industry/manufacturing/manufacturing-industry-outlook.html

[28] Murthy, G., & Anand, M. (2023, July 10). The road ahead for platform gig workers’ financial inclusion. CGAP. https://www.cgap.org/blog/road-ahead-for-platform-gig-workers-financial-inclusion

[29] Doshi, A. R., & Oliver, P. (2024). Hauser, Generative AI enhances individual creativity but reduces the collective diversity of novel content. *Science Advances*, *10*, eadn5290. doi: 10.1126/sciadv.adn5290

[30] Haywood, E. R. (2024). Harvesting Innovation: Exploring the Potential of an AI-Enabled Platform to Revolutionize Agricultural Labor Markets. Diss. Massachusetts Institute of Technology.

[31] Bagri, S., et al. (2025). “Empowering farmers and rural communities by harnessing ai for sustainable livelihoods.”. In *Integrating Agriculture, Green Marketing Strategies, and Artificial Intelligence* (pp. 163–180). IGI Global Scientific Publishing.

[32] Ali, Z., et al. (2025). “Artificial intelligence for sustainable agriculture: A comprehensive review of AI-driven technologies in crop production.”. *Sustainability*, *17*(5), 2281.

[33] Kesavan, V., & Srinivasan, K. S. “The role of artificial intelligence in sustainable development in industry revolution 4.0 economy. In *Synergy of AI and Fintech in the Digital Gig Economy* (pp. 386–416). CRC Press.

[34] Agrawal, K., & Kumar, N. (2025). “AI-ML applications in agriculture and food processing. In *Food and Industry 5.0: Transforming the Food System for a Sustainable Future* (pp. 21–37). Cham: Springer Nature Switzerland.

[35] Halubanza, B. (2024). *A framework for an early warning system for the management of the spread of locust invasion based on artificial intelligence technologies*. Diss. The University of Zambia.

[36] Patil, D. (2024). “Impact of artificial intelligence on employment and workforce development: Risks, opportunities, and socioeconomic implications. *Opportunities, And Socioeconomic Implications. (November 12, 2024)*.

Vasu Thakur, Vikas Kumar Roy, Ankush Mittal, Nupur Goyal*

Chapter 5
Analysis of Lockdown Effectiveness on Covid-19 Epidemic in Specific States of India by Using Artificial RNN Architecture

Abstract: The Covid-19 virus affected the world severely. It was declared a pandemic by WHO, and it drastically changed the lifestyle and thought processes of people. The loss of life, job crises, and adverse economic conditions occured worldwide. In developing countries like India, which has a large population, controlling the situation was very difficult. The most important step was imposing lockdowns in the first and second waves of Covid-19. In this chapter, we have attempted to predict the number of Covid-19 cases after imposing lockdowns during the second wave of the virus in Uttarakhand, Delhi, Himachal Pradesh, Goa, and Maharashtra. Here, we analyzed the impact of the lockdown on the spread of virus in the following states and how successful this lockdown was during second wave. We have used a deep learning algorithm, the long short-term memory (LSTM) model, which is a special type of artificial recurrent neural network architecture, to predict Covid-19 cases. We achieved almost the same results as expected, shown by the AUC, the predicted cases were almost the same as the confirmed cases.

Keywords: Covid-19, lockdown, recurrent neural network, LSTM model, area under curve

5.1 Introduction

Coronavirus disease (Covid-19) is an infectious disease caused by acute respiratory syndrome coronavirus 2 (SARS-CoV-2). In December 2019, the first known case of Covid-19 was identified in Wuhan, China. This virus spread all over the world, leading

***Corresponding author: Nupur Goyal,** Department of Mathematics, Graphic Era (Deemed to be University), Dehradun, Uttarakhand, India, e-mail: nupurgoyalgeu@gmail.com
Vasu Thakur, Department of Data and Computer Science, Ruprecht Karl University of Heidelberg, Germany, e-mail: vasuthakurin14@gmail.com
Vikas Kumar Roy, Department of Data and Computer Science, Ruprecht Karl University of Heidelberg, Germany, e-mail: royvikas1610@gmail.com
Ankush Mittal, Department of Data and Computer Science, College of Engineering Roorkee, Roorkee, Uttarakhand, India, e-mail: dr.ankushmittal@gmail.com

 | https://doi.org/10.1515/9783112211991-005

to a pandemic. A total of 20.4 crore cases were identified as of August 11, 2021. Total deaths caused by this disease are 43.1 lakhs worldwide, with 4.29 lakhs deaths recorded in India as of that date [19]. The symptoms of Covid-19 include fever, shortness of breath, loss of smell or taste, dry cough, and fatigue. Some cases are asymptomatic. Potential complications include kidney failure, respiratory failure, and other conditions. This disease is dangerous to every age group, but older people are more likely to be affected. Coronavirus transmits mainly through contact with people in close proximity. A person can become infected when droplets containing the virus are inhaled or come into direct contact with the eyes, nose, or mouth.

To control the transmission of this virus, a complete lockdown was imposed in many countries, including India. The impact of the lockdown affected the global economy significantly. During the first wave, the central government imposed a nationwide lockdown under the Disaster Management Act 2005. However, during the second wave, state governments imposed lockdowns based on their local Covid-19 cases. There were worst conditions, Indians lost employment, and the economy was devastated. The index of 8 core industries fell by 38.6% in April. Three months later, in June 2020, the index was still 15% lower than it was a year ago. Professional forecasters' estimates compiled by the Reserve Bank of India (RBI) during late July to early August 2020 showed that the real GDP was expected to shrink 22.8% in the quarter ended June 2020 compared to the corresponding quarter a year ago [1].

Having control over Covid-19 cases and providing the necessary facilities to patients, such as beds, medicine, oxygen, and doctors, is vital for avoiding burdens on the healthcare system. It is very important to predict the number of Covid-19 cases in the upcoming weeks and to analyze the impact of lockdowns on the spread of cases. If the prediction of Covid-19 cases and analysis of lockdown impacts on cases are done, it will help the government know when to impose the lockdowns in states and how many days it will take to control the spread of the virus.

Various implementations of machine learning algorithms for Covid-19 has been proposed, such as diagnosis and prognosis, patient outcome prediction, tracking and predicting the outbreak, drug development, vaccine discovery, and false news prediction [2, 3]. Many machine learning models have been used to predict the number of confirmed cases of Covid-19 [20; 4, 5]. The evolution of machine learning and deep learning has led to a great focus on predicting future data from previous ones. The most frequently used deep learning model is the recurrent neural network for the temporal problems. Deep learning is a subset of advanced machine learning methods, and is essentially a neural network with three or more layers that work like a human brain [6, 7].

Pham et al. [5] tried to emphasize the importance of advanced machine learning applications in responding to the Covid-19 outbreak and preventing its severe effects. They also provided an overview of AI and big data. They also identified their applications in fighting against Covid-19, the challenges and issues associated with state-of-the-art solutions, and gave recommendations for communication strategies to effectively control the Covid-19 situation. Bullock et al. [4] presented an overview of recent studies using machine learning and, more broadly, artificial intelligence to tackle the aspects of the Covid-19 crisis, such as identifying applications that address the challenges posed by Covid-19 at different scales. The authors also highlighted the need for international cooperation to maximize the potential of AI in this and future pandemics. Seven significant applications of AI for the Covid-19 pandemic have been identified and explained, how this technology plays an important role in detecting the cluster of cases and predicting where this virus will affect in future by collecting and analyzing all previous data. An algorithm based on RNN is designed in which the water search technique acquired by smart elephants [6]. Khan and Gupta [7] applied the univariate time series model to predict the number of Covid-19 infected cases that can be expected in upcoming days in India and adopted an auto-regressive integrated moving average (ARIMA) model on the data collected from January 31, 2020, to March 25, 2020, and verified it using the data collected from March 26, 2020, to April 4, 2020. A nonlinear autoregressive (NAR) neural network was developed to compare the accuracy of predicted models [8].

In this chapter, we have predicted the number of confirmed Covid-19 cases after imposing the lockdown and analyzed how effective the lockdown was in the second wave in the following five states of India (Maharashtra, Delhi, Uttarakhand, Himachal Pradesh, and Goa). In this study, we have used the artificial recurrent neural network architecture, LSTM (long short-term memory), for forecasting the number of Covid-19 cases after imposing the lockdown in these five states of India. The LSTM is trained on the data prepared by WHO, and the prediction of Covid-19 cases is made. The authors received almost the same forecast as the actual number of Covid-19 cases as verified. The reason behind obtaining the right prediction is the algorithm that was used.

The chapter is organized as follows: long short-term memory model (Section 5.2), data analysis (Section 5.3), results and discussion (Section 5.4), and conclusion (Section 5.5).

5.2 Long Short-Term Memory (LSTM) Model

LSTM is a special type of artificial recurrent neural network (RNN) architecture used in the field of deep learning [9]. The intuition acquired by analyzing the RNN applies to LSTM too [10]. LSTM is used to predict future data after being trained on available data because it is suitable for forecasting time series sequence. The error rates of LSTM are substantially less than the other methods [11]. It was introduced by Hochreiter and Schmidhuber in 1997. These networks were built to get rid of long-term dependency problems. LSTM can delete or add data to the cell, which is structured with gates. Gates are a method to allow the data through them. Various studies have used the LSTM model to forecast future data and have achieved high accuracy with nearly identical predictions. The reason behind using this model is its architectural design. This model consists of an input layer, multiple hidden layers, and an output layer. The memory cell in the hidden layer is a key characteristic of this model, and data are learned by maintaining the status of the memory cell [12]. The comprehensive architecture of LSTM is shown in Figure 5.1.

LSTM has "cells" in the hidden layers of the neural network, which contain three gates – an input gate, an output gate, and a forget gate. These gates control the flow of information required to predict the network's output [9, 13] and are used to prevent from vanishing gradient problem [14]. These gate functions are needed to achieve long-term dependency problem due to its distinctive storage unit architecture [15, 16].

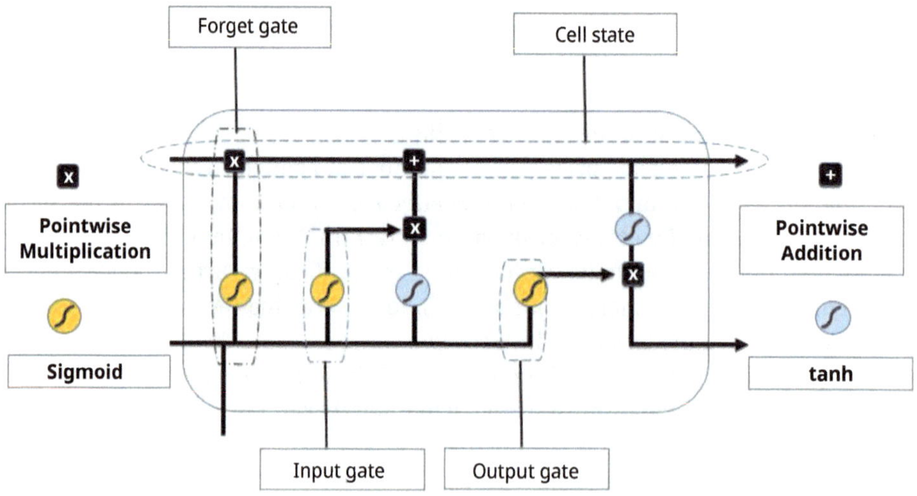

Figure 5.1: LSTM architecture.

The gates sort through sequences of data, retaining useful information and removing irrelevant data. Due to this, the relevant data passes through these sequential chains to forecast trends. The major applications of LSTMs include: speech recognition, video caption generation, text generation, machine learning, and coding and decoding technology [17]. LSTM, an artificial recurrent neural network architecture is better than rule-based algorithms [18] and is generally used for time series forecasting.

5.3 Data Analysis

The most accessible problem for the RNN model is to solve the most challenging task like forecasting future data. To achieve accurate predictions, we need reliable data in the correct structure. The most difficult task in predicting Covid-19 cases is data collection. The shortage of data made it difficult for the authors to predict the infected cases. The dataset was collected manually from the WHO and was organized in an incorrect structure.

The Covid-19 case data for Maharashtra, Delhi, Uttarakhand, Himachal Pradesh, and Goa are represented in bar graphs as shown in Figures 5.2–5.6. Table 5.1 lists the states, first date, and last date on which the lockdown was imposed, the number of Covid-19 cases on those days, and the number of days for which the dataset was considered.

Table 5.1: Data of Covid-19 cases in five states.

States	First date	Covid-19 cases on the first date	Last date	Covid-19 cases on the previous date	Number of days
Uttarakhand	11/05/2021	7,120	09/06/2021	513	30
Delhi	19/04/2021	23,686	02/06/2021	576	45
Himachal Pradesh	07/05/2021	4,177	05/06/2021	995	30
Goa	29/04/2021	3,019	28/05/2021	1,055	30
Maharashtra	14/04/2021	58,952	28/05/2021	20,740	45

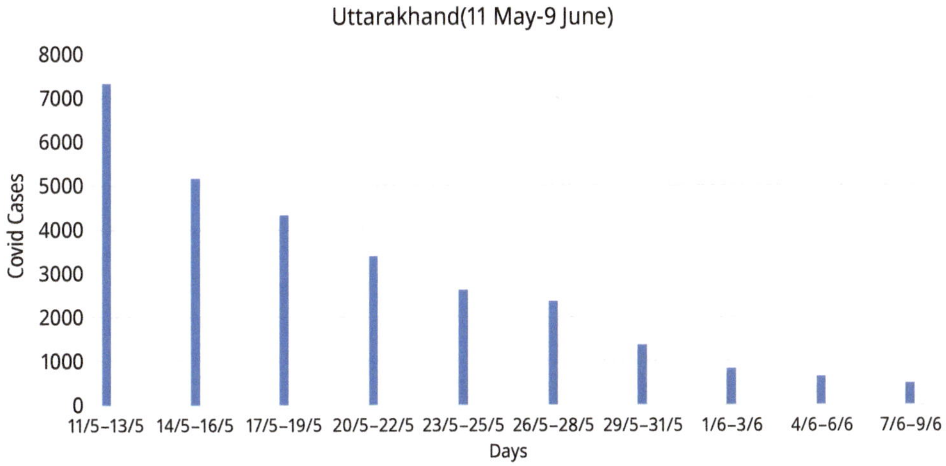

Figure 5.2: Data of Covid-19 cases in Uttarakhand.

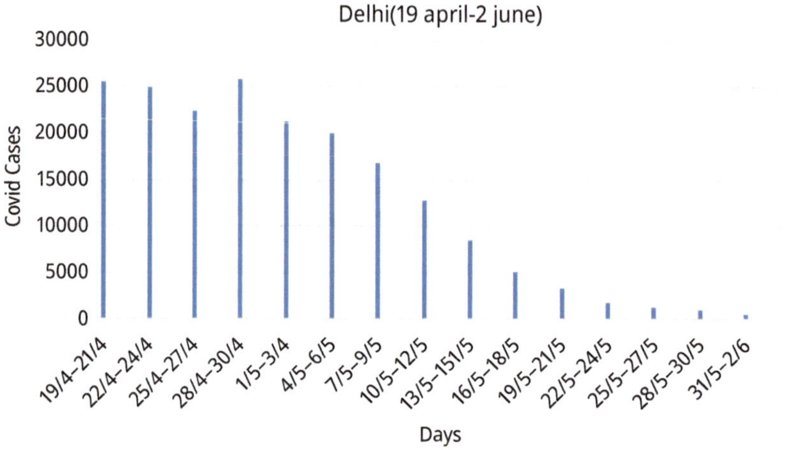

Figure 5.3: Data of Covid-19 cases in five Delhi.

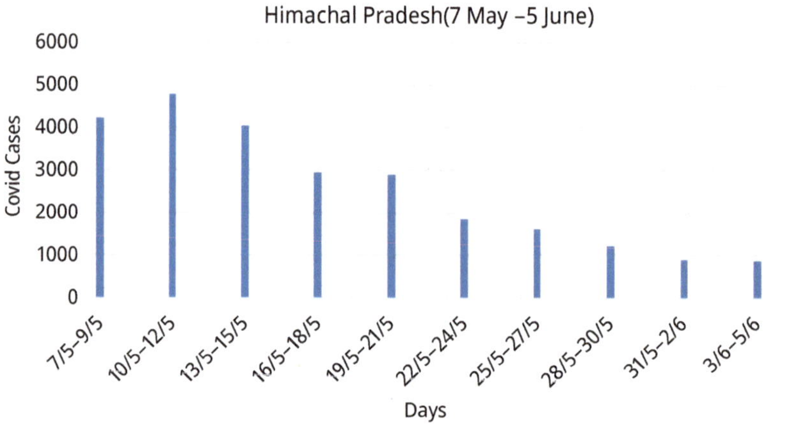

Figure 5.4: Data of Covid-19 cases in Himachal Pradesh.

5.3.1 Data of Covid-19 Cases in Five States

We considered the average of three days of Covid-19 cases on the Y-axis and the corresponding dates on the X-axis. Different state governments imposed lockdowns on different dates. Therefore, we have also mentioned the other starting dates. Our dataset is taken from one week before the week in which the state government imposed a lockdown, for the purpose of training and predicting the model. The comparative study of Covid-19 cases for the five states is presented by area under curve (AUC), as revealed in Figure 5.7 (number of days on the X-axis and number of Covid-19 cases on the Y-axis).

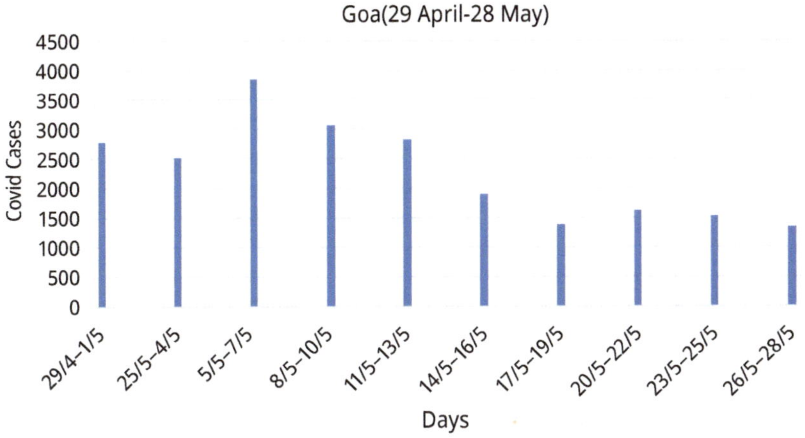

Figure 5.5: Data of Covid-19 cases in Goa.

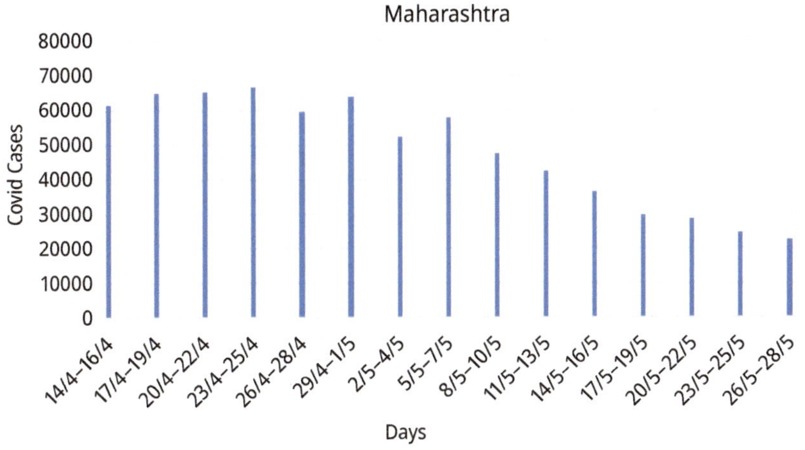

Figure 5.6: Data of Covid-19 cases in Maharashtra.

The LSTM model was designed, as described by the flow chart shown in Figure 5.8, to predict the number of Covid-19 cases after imposing lockdowns in Maharashtra, Delhi, Uttarakhand, Himachal Pradesh, and Goa. The results were almost as expected; achieving almost the same accuracy as the pattern of a graph of predicted results matched with the actual one. The authors have also analyzed the per capita Covid-19 cases and normalized them to get better results. The flow diagram for per capita Covid-19 cases is illustrated in Figure 5.9. Mathematically, it can be calculated as:

$$\text{Per capita tests} = \frac{\text{total number of tests}}{\text{number of people in that specific area}}$$

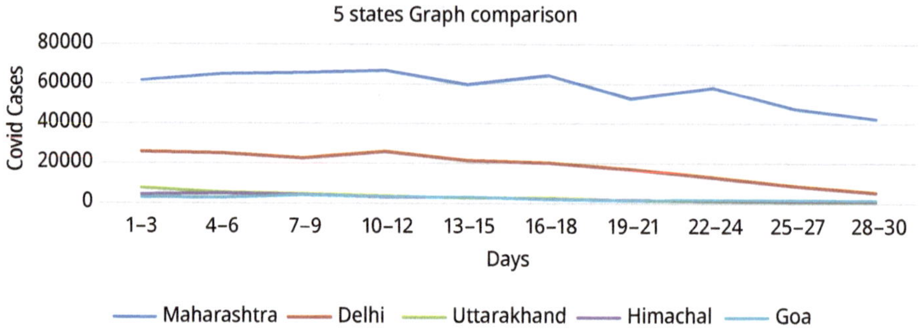

Figure 5.7: Comparative study of Covid-19 cases.

Example:

The population of Goa is 1.59 million. The average of first three days' Covid-19 cases in Goa = (582 + 428 + 540)/3 = 516.667 Covid-19 cases. Calculating 516.667 cases for a population of 100,000 results in 32.4948 cases. Hence, per capita cases are 32.4948.

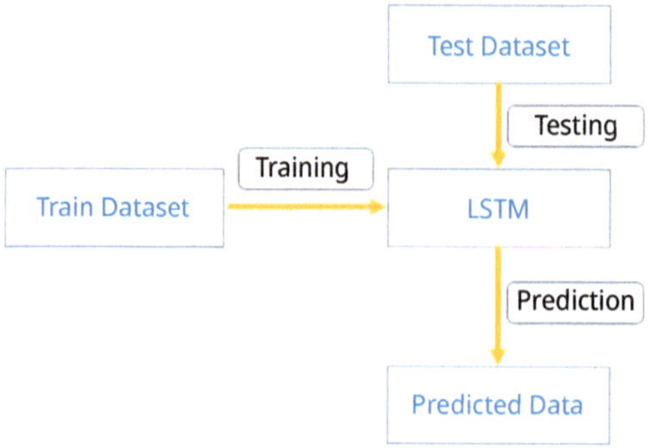

Figure 5.8: LSTM flow diagram.

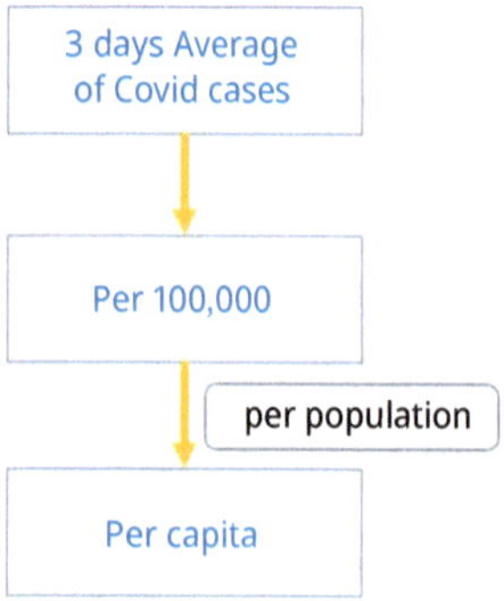

Figure 5.9: Per capita flow diagram.

5.4 Results and Discussion

The LSTM model has tremendous potential to predict future data with high accuracy. The computational procedure for the collected data is carried out through time-series analysis, and is done with good accuracy.

In the first scenario, the authors analyze how the lockdown affects the number of Covid-19 cases in these states. The lockdown had a very positive effect on controlling the spread of corona virus. The decision to impose the lockdown was very right because, as shown in Figure 5.10, the per capita Covid-19 cases decreased rapidly after the first day of the lockdown.

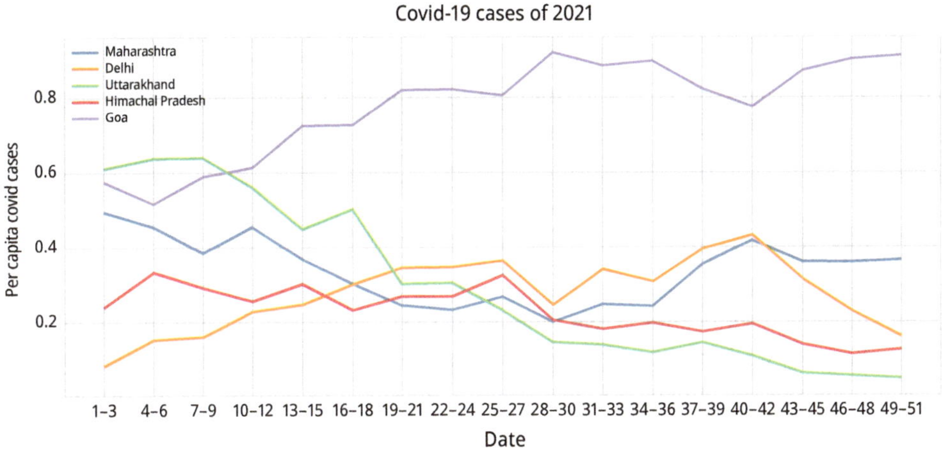

Figure 5.10: Graphical representation of per capita Covid-19 cases.

Initially, when the lockdown was imposed, the cases were increasing. As the number of days of lockdown increased, there is a decrease in Covid-19 cases, which shows the successful implementation of the lockdown model. However this was not fully successful everywhere, as shown in the case of Goa. This may be because of the limitations of the dataset used. However, overall, the count of Covid-19 cases decreased in these states.

In scenario 2, we are predicting the number of Covid-19 cases per capita after imposing the lockdown and comparing them with the actual cases recorded in these states, as shown in Figure 5.11.

The graphical representation of the area under curve (AUC) of these states is mentioned.

The prediction of Covid-19 cases is made after the 16th day of lockdown. We achieved the same output as expected, as can be seen in this graph. The predicted and

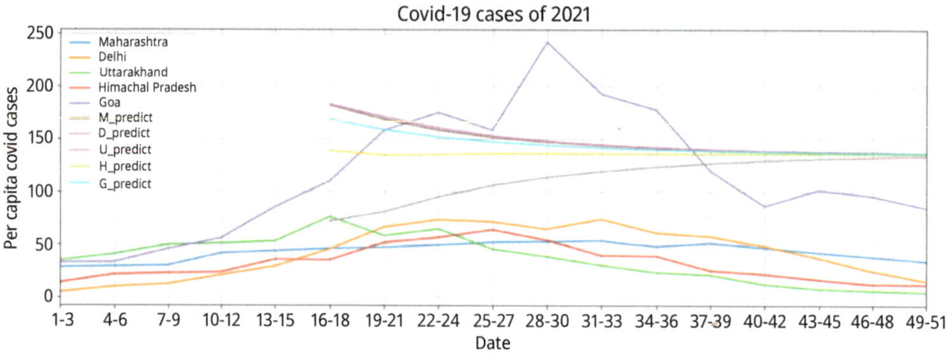

Figure 5.11: Graphical representation of prediction of Covid-19 cases.

solid lines decrease, indicating that Covid-19 cases are rapidly falling after imposing the lockdown in states like Uttarakhand, Maharashtra, Himachal Pradesh, and Delhi.

5.5 Challenges

After imposing the lockdown, we have used LSTM models to predict the number of Covid-19 cases and achieved almost the same falling graph for predicted Covid-19 cases compared to confirmed cases. Still, the authors faced many difficulties and encountered some failure cases for the following reasons:

- Lack of huge amount of historical datasets because of the limitation of the date, and getting the exact number of confirmed cases on the following date is very difficult. Therefore, our model did not train exactly because the deep learning model needs a huge dataset to achieve successful results.
- Our models have failed in the case of Goa because, as can be seen in the graph of AUC in Figure 5.6, Covid-19 cases increase even after imposing the lockdown. Possible reasons for these failure cases may be that we have taken our dataset only for short periods. Different states had different stages of Covid-19 cases, and the various governments imposed additional rules and liabilities in these states. Most importantly, the population density of these states is also different. These may be the reasons for failure cases.

5.6 Conclusion

In this chapter, the authors have analyzed the effectiveness of lockdowns when the second wave of Covid-19 epidemic started in Uttarakhand, Delhi, Himachal Pradesh, Goa, and Maharashtra for a certain period. This analysis helps determine how effective the lockdown was at controlling the viruses and the losses common people suffered because the lockdown was worthwhile in the context of life-saving measures. Most importantly, this will help the government know when to impose a lockdown in states to control the spread of Covid-19 if there are any further waves. It will also show how many days it will take to bring it under control and will also help estimate how many possible cases will be encountered after certain days of imposing the lockdown. This will help the government make possible arrangements for hospitals, beds, doctors, oxygen, medicine, and health workers for upcoming Covid-19 cases. Not only this, but if the government knows how many cases will be faced in the upcoming future, they will be able to prepare the future requirements and decisions regarding liability in future lockdowns will be taken accordingly so that people and the government will face minimum losses.

Now, in India, vaccination is on a large scale. This will be important for controlling the Covid-19 virus. Therefore, it is important to know the proper speed of vaccination for controlling the third wave. Therefore, our next work is to predict the number of Covid-19 cases after the successful vaccination in India.

References

[1] Vyas, M. (2020). Impact of lockdown on labour in India. *The Indian Journal of Labour Economics, 63*(1), 73–77.

[2] McCall, B. (2020). COVID-19 and artificial intelligence: Protecting health-care workers and curbing the spread. *The Lancet Digital Health, 2*(4), e166–e167.

[3] Vaishya, R., Javaid, M., Khan, I. H., & Haleem, A. (2020). Artificial Intelligence (AI) applications for COVID-19 pandemic. *Diabetes and Metabolic Syndrome: Clinical Research and Reviews, 14*(4), 337–339.

[4] Bullock, J., Luccioni, A., Pham, K. H., Lam, C. S. N., & Luengo-Oroz, M. (2020). Mapping the landscape of artificial intelligence applications against COVID-19. *Journal of Artificial Intelligence Research, 69*, 807–845.

[5] Pham, Q. V., Nguyen, D. C., Huynh-The, T., Hwang, W. J., & Pathirana, P. N. (2020). Artificial intelligence (AI) and big data for coronavirus (COVID-19) pandemic: A survey on the state-of-the-arts. *IEEE Access, 8*, 130820.

[6] Mandal, S., Saha, G., & Pal, R. K. (2017). Recurrent neural network-based modeling of gene regulatory network using elephant swarm water search algorithm. *Journal of Bioinformatics and Computational Biology, 15*(04), 1750016.

[7] Khan, F. M., & Gupta, R. (2020). ARIMA and NAR based prediction model for time series analysis of COVID-19 cases in India. *Journal of Safety Science and Resilience, 1*(1), 12–18.

[8] Benuwa, B. B., Zhan, Y. Z., Ghansah, B., Wornyo, D. K., & Kataka, F. B. (2016). A review of deep machine learning. *International Journal of Engineering Research in Africa, 24*, 124–136.

[9] Hochreiter, S., & Schmidhuber, J. (1997). Long short-term memory. *Neural Computation, 9*(8), 1735–1780.

[10] Sherstinsky, A. (2020). Fundamentals of recurrent neural network (RNN) and long short-term memory (LSTM) network. *Physica D Nonlinear Phenomena, 404*, 132306.

[11] Liu, T., Wu, T., Wang, M., Fu, M., Kang, J., & Zhang, H. (2018, September). Recurrent neural networks based on LSTM for predicting geomagnetic field. In *2018 IEEE International Conference on Aerospace Electronics and Remote Sensing Technology (ICARES)* (pp. 1–5). IEEE.
[12] Jeon, B. K., & Kim, E. J. (2021). LSTM-based model predictive control for optimal temperature set-point planning. *Sustainability*, *13*(2), 894.
[13] Education, I. C. (2020). Recurrent neural networks. https://www.ibm.com/cloud/learn/recurrent-neural-networks.
[14] Wang, Y. (2017, May). A new concept using lstm neural networks for dynamic system identification. In *2017 American Control Conference (ACC)* (pp. 5324–5329). IEEE.
[15] Yu, Y., Si, X., Hu, C., & Zhang, J. (2019). A review of recurrent neural networks: LSTM cells and network architectures. *Neural Computation*, *31*(7), 1235–1270.
[16] Qiu, J., Wang, B., & Zhou, C. (2020). Forecasting stock prices with long-short term memory neural network based on attention mechanism. *PloS One*, *15*(1), e0227222.
[17] Wang, C., Du, W., Zhu, Z., & Yue, Z. (2020). The real-time big data processing method based on LSTM or GRU for the smart job shop production process. *Journal of Algorithms and Computational Technology*, *14*, 1748302620962390.
[18] Leevy, J. L., Khoshgoftaar, T. M., & Villanustre, F. (2020). Survey on RNN and CRF models for de-identification of medical free text. *Journal of Big Data*, *7*(1), 1–22.
[19] https://www.who.int/countries/ind.
[20] Wim, N. (2020). *Artificial Intelligence against Covid-19: An Early Review. IZA DP, (13110), Institute of Labor Economics (IZA)*: Bonn.

Garvita Arora, Akansha Gupta

Chapter 6
Computational Approaches to Epilepsy Detection: ML and DL Perspectives

Abstract: Epilepsy is defined as a neurological disorder that is caused by aberrant neuronal firing in the brain. Early and accurate diagnosis of epilepsy plays an important role in improving treatment plans for patients, which further helps in reducing complications and improving patients' quality of life. However, traditional approaches often face difficulties such as manual electroencephalography (EEG) interpretation, requiring expert dependency, challenges in handling large EEG data, and variability in seizure patterns. This review presents an outline of epilepsy diagnosis by studying conventional neurodiagnostic techniques such as EEG, magnetic resonance imaging, and other imaging modalities, along with different phases of seizures. Furthermore, the study reviews the automation of early diagnosis and seizure detection by integrating machine learning and deep learning techniques, which help in pattern recognition from EEG signals, thereby reducing manual effort, improving diagnostic reliability, and facilitating personalized neurological treatment strategies. In addition, for better comparative analysis and evaluation, some commonly used benchmark datasets are discussed. By integrating neurodiagnostic techniques with artificial intelligence (AI)-based techniques, researchers aim to improve the detection of early seizures. With the ultimate goal of transforming epilepsy detection and diagnosis into a more fair and successful paradigm for better patient outcomes globally, future directions support explainable AI, portable diagnostics, and ethical frameworks to address data privacy and accessibility discrepancies.

Keywords: Epilepsy, epileptic seizure detection, electroencephalography (EEG), seizure detection, machine learning, deep learning, artificial intelligence, EEG signal processing

6.1 Introduction

Epilepsy is a brain disease that affects people through repeated seizures caused by abnormal brain activity [1]. The signs of this disease can range from mild disturbances to severe convulsions [2]. This is why it isn't easy to diagnose and treat. This chap-

Garvita Arora, Department of Computer Science, Graphic Era (Deemed to be University), Dehradun, Uttarakhand, India, e-mail: garvitaarora777@gmail.com
Akansha Gupta, Department of Computer Science, Graphic Era (Deemed to be University), Dehradun, Uttarakhand, India, e-mail: akanshagupta.cse@geu.ac.in

 | https://doi.org/10.1515/9783112211991-006

ter provides a review of the symptoms of epilepsy, its clinical value, its detection challenges, and the transformation of combining conventional methods with new artificial intelligence (AI)-based detection techniques [3]. It focuses on how to improve health outcomes and quality of life by focusing on worldwide effects, the necessity of early detection, and advanced technologies [4].

6.1.1 Overview of Epilepsy

A long-standing brain condition called epilepsy is characterized by a tendency for recurrent seizures that don't have an immediate cause. Seizures happen when a group of brain cells suddenly sends out too much electrical activity, and this disturbs the brain's normal functioning [1, 2]. Different people experience seizures in different ways. Some may stare blankly for a few seconds, and some may shake violently and lose consciousness [2].

Doctors now understand that epilepsy is not just a single disease. It is a group of different disorders that have many causes. Sometimes seizures occur after a brain injury such as a tumor, brain infection, stroke, or severe head trauma. However, in about one out of three people, doctors cannot find the exact cause, even after many tests [3].

Epilepsy affects more than just the seizures themselves. Frequent seizures can affect learning, memory, and emotions over time. People with epilepsy may face misunderstanding from others, which can make it harder to drive, get jobs, or maintain relationships. There is a small but real risk of getting injured during a seizure or of sudden unexpected death [1, 3].

To diagnose epilepsy, doctors first listen to the patient carefully. Then they may use a test called an electroencephalography (EEG) to check for unusual brain wave patterns. Brain scans are also used to see structural problems. Treatment usually begins with medication to control brain activity. If medicines do not work, then surgery may be considered. Even though treatments are available, many challenges remain where medical care is limited. Treatment usually starts with medication to stabilize brain activity after confirmation, but for patients who don't respond well to medication, surgery or other methods may be taken into consideration [2, 3].

Even with effective treatment choices, there are still numerous obstacles to overcome, particularly in regions of the world with few medical resources. Ongoing advancements in the detection and understanding of this disorder are crucial, since an early and correct diagnosis is essential for improved control and fewer long-term issues [4, 23].

6.1.2 Global Burden and Statistics

Epilepsy accounts for over 13 million disability-adjusted life years (DALYs) and is responsible for more than 0.5% of the global burden of disease (GBD). It affects people of all ages, sexes, races, income groups, and geographical locations. Around 7.6 per 1,000 persons have epilepsy during their lifetime. It has a bimodal distribution according to age, with peaks in the youngest individuals and in those over 60 years of age. According to disability-adjusted life years (DALYs), which combine years lost to disability and premature death, epilepsy is responsible for almost 0.5% of the worldwide burden of illness [5]. Nearly 80% of the burden falls on low- and middle-income nations, where treatment access is scarce and treatment gaps can reach 75% [4]. In 2021–2022, around 2.9 million adults in the United States had active epilepsy, which accounted for 1% of the adult population [7]. In 2021, epilepsy of unknown cause alone resulted in millions of cases of morbidity and fatality worldwide [5]. Although overall trends point to a rising prevalence as a result of aging populations and better survival from causative illnesses like strokes, projections show a drop in the burden of pediatric epilepsy by 2035, particularly in children under 5 [6].

Beyond just health, this burden includes stigma, fewer prospects for education and employment, and higher healthcare costs. Untreated epilepsy increases the risk of accidents, drowning, and sudden unexpected death in epilepsy (SUDEP) in low-resource settings, while its annual expenditures may surpass $10,000 per patient in high-income nations [7, 8].

6.1.3 Importance of Early Diagnosis

Early diagnosis is important for avoiding complications, enhancing the quality of life, and maximizing the best possible patient outcome. The prompt administration of AEDs can lessen the likelihood of injury, impairing cognition, and psychosocial issues through the control of seizures in as high a percentage as 70% of all cases [9]. Critical periods of development are when seizures need to be identified and treated early because, if left untreated, they could lead to behavioral issues or learning disabilities in children [10]. In infancy, unmonitored seizures can result in greater damage to the brain than in an infant who has been diagnosed and monitored early on [10].

Research highlights that a worse prognosis is associated with delayed diagnosis; waiting periods following the initial seizure might affect the course of epilepsy and the effectiveness of treatment [11]. In order to guide individualized care and prevent needless exposure to ineffective medications, an early examination makes it possible to classify seizure types and etiologies [11]. It also facilitates easier access to support programs and lifestyle changes, such as avoiding triggers. Early treatment prevents prolonged discharges in conditions like continuous spike-wave during sleep (CSWS), which improves cognitive outcomes [10]. Early diagnosis plays a crucial role in turning epilepsy from a crippling condition to one that can be managed, as it lowers the socioeconomic burden by reducing hospitalizations and lost productivity [9–11].

6.1.4 Difficulties in the Present Diagnosis

Diagnosing epilepsy remains challenging due to various technical and clinical factors. Seizures occur for very short durations, making real-time observation difficult. Due to this, diagnosis often relies on reports from family members or the patient's medical history, which may be inaccurate or incomplete at times. In many cases, epilepsy is wrongly diagnosed as a mental disorder, syncope, or nonepileptic events, which leads to a delay in proper treatment [12].

In areas where there are low-resource settings or limited accessibility to advanced diagnostic technologies, complications in accurate epilepsy detection are further worsened [4]. The lack of usage of video-EEG monitoring techniques in many areas is restricting comprehensive evaluation. Social stigma and misconceptions surrounding epilepsy also create a lack of timely medical attention. In addition, other conditions like depression and anxiety can make diagnosis more difficult and complex [13].

Technically, conventional EEG recordings may not easily detect seizures originating deep inside the brain, often leading to long-term monitoring to capture infrequent seizure events [14]. Integrating different types of medical data is difficult due to the presence of signal noise, variability among patients, and inconsistencies in data acquisition. Other factors, such as differences in epilepsy etiology and abnormal EEG findings, often lead to variability in clinical outcomes. Addressing these challenges requires the development of advanced diagnostic techniques and AI-based analysis to improve accessibility and accuracy.

6.2 Types of Epilepsy

The International League Against Epilepsy (ILAE) [16] has categorized epilepsy into different types based on its causes and features, as described in Figure 6.1. The four main types of epilepsy include:

- **Focal epilepsy**: Seizures start in one specific area of the brain.
- **Generalized epilepsy**: Seizures begin on both sides of the brain at the same time.
- **Combined generalized and focal epilepsy**: A person has both focal and generalized seizures.
- **Unknown epilepsy**: This type is not clear because there is not enough information.

Table I: Types of Epilepsy

Epilepsy Type	Description	Common Seizure Types
Focal Epilepsy	Seizures originate in one area or network of brain cells on one side of the brain.	Focal aware, focal impaired awareness, focal to bilateral tonic-clonic.
Generalized Epilepsy	Seizures involve networks on both sides of the brain from the start.	Absence, myoclonic, tonic-clonic, atonic.
Combined Generalized and Focal Epilepsy	Person experiences both focal and generalized seizures.	Mixture of above.
Unknown Epilepsy	Onset cannot be determined as focal or generalized (e.g., due to limited information).	Variable, often tonic-clonic or unknown onset.

The table explains each type and kind of seizure to make the diagnosis more accurate. Doctors are now using genetic testing and brain imaging. Imaging scans like magnetic resonance imaging help classify epilepsy more clearly [17–20].

6.3 Neurodiagnostic Techniques

To find out where seizures start in the brain, what might be causing them, and to confirm if a person has epilepsy, doctors use special tests like detailed pictures of the brain structure and functions, while others record the brain's electrical activity.

One of the most important tests is the EEG. In this test, electrodes are placed on the scalp to record brain electrical signals, and it is very helpful in finding abnormal patterns, such as sharp waves, which appear during seizures and between seizures. Sometimes, doctors use video-EEG monitoring. This helps doctors match changes in brain waves with what the patient is physically experiencing [14]. A routine EEG can detect seizure-related changes in about 50% of the cases; longer monitoring increases the likelihood of finding abnormalities.

6.4 EEG Electrode Placement

To ensure EEG results are accurate and doctors follow a standard method for placing electrodes, the most commonly used method is the 10–20 system developed by the International Federation of Clinical Neurophysiology (IFCN). In this system, 21 electro-

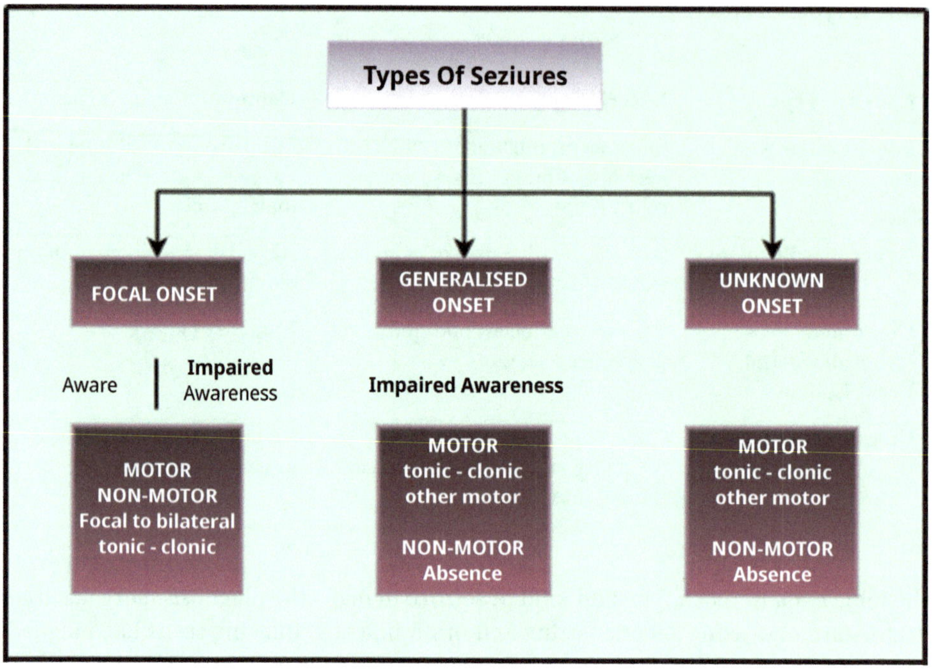

Figure 6.1: Types of epilepsy.

des are placed on the scalp, and their positions are measured at 10% or 20% distances between skull landmarks like the nasion (bridge of the nose), the inion (back of the head), and preauricular sites [21].

Each electrode has a name based on the brain lobe (F: frontal, T: temporal, C: central, P: parietal, O: occipital) and hemisphere (odd: left, even: right, z: midline) according to where the electrodes are placed. For example, Fp1/Fp2 and the center midline of Cz are frontopolar [21]. This setup helps doctors identify differences between the left and the right sides of the brain, which is important in diagnosing focal epilepsy. There is an advanced version called the 10–10 system, which uses more electrodes (up to 64) and provides more detailed brain activity, which is useful in epilepsy monitoring units.

Seizures are best detected using frontotemporal electrodes (such as F7-T5) [14]. In order to lower impedance, conductive paste and scalp abrasion are used in preparation. Artifacts from muscle or movement provide difficulties, which are lessened by bipolar or referential montages [14]. The diagnosis of epilepsy is supported by this system, which makes source localization possible through methods such as electrical source imaging (ESI) [21].

6.5 Epilepsy Seizure Phases

To help with classification and treatment, seizure phases show how epileptic occurrences change over time. The prodrome (hours to days of nebulous symptoms, such as irritation) and aura (seconds before, focal sensory abnormalities) are also included in the pre-ictal phase [22]. The seizure proper, known as the ictal phase, lasts anywhere from seconds to minutes and is characterized by clinical signs of neuronal hypersynchrony [22].

There is a post-ictal phase that lasts for minutes to hours and is marked by recovery with disorientation, exhaustion, or focal deficiencies (Todd's paralysis) [22]. EEG spikes or background abnormalities without symptoms may be seen during inter-ictal periods, which are seizure-free times [22]. Pro-ictal states, for instance, allow forecasting since they reflect brain changes hours in advance [22].

6.6 Conventional Signal Processing Techniques

Traditional signal processing uses time-frequency analysis to extract EEG data for the purpose of detecting epilepsy. The effectiveness of the fourier transform (FT), which breaks down signals into frequency bands (delta: 1–4 Hz, theta: 4–8 Hz) to find rhythmic patterns in post-ictal or focal seizures, is limited by the presumption of signal stationarity [15]. This limitation is addressed by the wavelet transform (WT), which uses multi-resolution analysis to detect transients and epileptiform spikes using continuous (CWT) or discrete (DWT) forms. In alpha bands (8–13 Hz), WT can detect these spikes with up to 92.82% accuracy [15]. Short-time FT (STFT) creates spectrograms for temporal analysis, while empirical mode decomposition (EMD) separates intrinsic mode functions [15].

Preprocessing includes segmenting data and filtering noise (such as 50/60 Hz powerline interference), then extracting features (such as entropy and power spectral density) and classifying the data using statistical models [15]. Although these technologies allow real-time monitoring in epilepsy units, they still have some limitations. Traditional analytical techniques often require manual parameter tuning and expert dependency, which becomes a challenge when EEG signals exhibit temporal variability and patient-to-patient differences. These methods provide limited generalization capability due to dynamic brain activity. For instance, while selecting an appropriate wavelet function, such as the Daubechies wavelet, is essential for signal analysis, this selection is frequently subjective and based on expert decisions. Due to these limitations and challenges, AI-based methods are needed to handle complex EEG patterns to improve diagnostic accuracy while making the process more efficient and scalable [15].

6.7 Machine Learning and Deep Learning in Epilepsy Detection

Machine learning (ML) and deep learning (DL) have improved the way doctors diagnose epilepsy using EEG brain signals. Traditional learning methods, such as random forest and support vector machine (SVM), study EEG data using manually selected features like signal frequency and wave patterns to detect seizures, achieving 90% accuracy.

Convolutional neural networks (CNNs) and recurrent neural networks (RNNs) can analyze raw EEG signals directly without manually chosen features, and they work better with noisy data and can reach up to the 98% accuracy. CNNs are good at finding patterns across different EEG channels, while RNNs are good at finding patterns across different EEG channels. They are better at understanding low signals that change over time to help predict seizures. Some advanced CNN models, like entropy measures, are reliable and can achieve more than 90% and 95% sensitivity in children, especially. AI models in wearable devices to predict seizures up to 30 minutes before they happen, with 80% accuracy with real-time monitoring [23].

However, there are some challenges. Some AI models are difficult to understand and need powerful computing power, and may overfit the training data. New techniques such as explainable AI and improved neural networks like bidirectional LSTMs are helping to solve these problems.

Overall, AI is transforming epilepsy care by making diagnosis faster, more accurate, and more accessible, especially in areas with limited medical resources, while also reducing the workload for doctors.

6.8 Benchmark Datasets

Benchmarking datasets provide a foundation for developing and validating ML/DL algorithms to identify epilepsy. The CHB-MIT scalp EEG database at Children's Hospital Boston provides the continuous EEG (>900 h) of 23 pediatric patients who have experienced documented seizures; while there were 10 second delays in some sections of the dataset, it is an ideal resource for long-term studies [24].

The Bonn University dataset is comprised of five (A–E) groups of single-channel EEG recordings from both healthy individuals and those with epilepsy, utilizing high frequency sampling (173.61 Hz). The recordings have been utilized in the binary classification of seizures (versus nonseizure) [25].

The Temple University Hospital (TUH) EEG corpus is the largest publicly available dataset, providing over 30,000 adult sessions, including artifacts, which can be used as a training set for real-world noise [26].

Utilizing CHB-MIT and Bonn as benchmarks for ML/DL models allows for cross validation [23]. The use of private data for evaluation in current benchmarks promotes the development of generalizable AI systems [23].

6.9 Future Directions and Conclusion

Future research will utilize AI to help doctors diagnose and predict seizures more accurately. AI systems will combine brainwave data (EEG) with other information, such as heart rate, wearable device data, and video recordings [23]. This approach may allow seizures to be predicted with more than 90% accuracy. When a doctor works with AI, EEG results can be analyzed faster, which helps reduce delays in diagnosis. In areas with limited medical resources, portable EEG devices and cloud-based technology will allow doctors to monitor patients remotely through telemedicine [4]. However, challenges remain, including protecting patient data, ensuring AI works well for different populations, and using AI in an ethical way, especially to avoid bias and ensure equal access. To manage these challenges, governments and organizations are developing new regulations for AI in healthcare.

Overall, the transitions from traditional diagnostic approaches toward AI-driven systems that integrate multimodal medical data helps in improve detection and prediction accuracy. Furthermore, providing enhanced treatment planning contributes to global healthcare improvement [23].

References

[1] World Health Organization: WHO, "Epilepsy" (Feb. 07, 2024). https://www.who.int/news-room/fact-sheets/detail/epilepsy

[2] Fisher, R. S., et al. (Apr. 2014). ILAE official report: A practical clinical definition of epilepsy. *Epilepsia, 55*(4), 475–482. doi: 10.1111/epi.12550

[3] Shoeibi, A., et al. (May 2021). Epileptic seizures detection using deep learning techniques: A review. *International Journal of Environmental Research and Public Health, 18*(11), 5780. doi: 10.3390/ijerph18115780

[4] "Epilepsy: a public health imperative" (Jun. 13, 2019). https://www.who.int/publications/i/item/epilepsy-a-public-health-imperative

[5] Feigin, V. L., et al. 2021. Global, regional, and national burden of epilepsy, 1990–2021: A systematic analysis for the Global Burden of Disease Study. *The Lancet Public Health, 10*(3), e203–e227.

[6] Yang, L.-Z., Guo, Y., Wang, Z.-Q., & Zhang, C.-Q. (Dec. 2024). A population-based analysis of the global burden of epilepsy across all age groups (1990–2021): Utilizing the Global Burden of Disease 2021 data. *Frontiers in Neurology, 15*, 1448596. doi: 10.3389/fneur.2024.1448596

[7] Centers for Disease Control and Prevention. (2024). *Epilepsy Data and Statistics*. CDC, Atlanta, GA, USA.

[8] Beghi, E., Giussani, G., & Sander, J. W. (Sep. 2015). The natural history and prognosis of epilepsy. *Epileptic Disorders, 17*(3), 243–253. doi: 10.1684/epd.2015.0751

[9] "Early diagnosis, treatment and prognosis of epilepsy with continuous spikes and waves during slow sleep", *PubMed*. (Mar. 15, 2015). https://pubmed.ncbi.nlm.nih.gov/26064309/

[10] Khan, A., Lim, H., & Almubarak, S. (May 2020). Importance of prompt diagnosis in pediatric epilepsy outcomes. *Seizure*, *80*, 24–30. doi: 10.1016/j.seizure.2020.03.011

[11] Kwan, P., & Brodie, M. J. (Feb. 2000). Early identification of refractory epilepsy. *New England Journal of Medicine*, *342*(5), 314–319, doi 10.1056/nejm200002033420503

[12] Hampel, K. G., Garces-Sanchez, M., Gomez-Ibanez, A., Palanca-Camara, M., & Villanueva, V. (Jan. 2019). Desafíos diagnósticos en epilepsia. *Revista De Neurología*, *68*(06), 255. doi: 10.33588/rn.6806.2018242

[13] Ballini, E. E. (2023). *Syndromes and Seizures: Case Studies in Cognition and Mental Disorders of Adults with Epilepsy*. doi: 10.25903/gthw-3w04

[14] Ye, S., Bagić, A., & He, B. (Dec. 2023). Disentanglement of resting state brain networks for localizing epileptogenic zone in focal epilepsy. *Brain Topography*, *37*(1), 152–168. doi: 10.1007/s10548-023-01025-z

[15] Amin, H. U., Yusoff, M. Z., & Ahmad, R. F. (Nov. 2019). A novel approach based on wavelet analysis and arithmetic coding for automated detection and diagnosis of epileptic seizure in EEG signals using machine learning techniques. *Biomedical Signal Processing and Control*, *56*, 101707. doi: 10.1016/j.bspc.2019.101707

[16] Fisher, R. S., et al. (Mar. 2017). Operational classification of seizure types by the International League Against Epilepsy: Position Paper of the ILAE Commission for Classification and Terminology. *Epilepsia*, *58*(4), 522–530. doi: 10.1111/epi.13670

[17] Beniczky, S., et al. (Apr. 2025). Updated classification of epileptic seizures: Position paper of the International League Against Epilepsy. *Epilepsia*, *66*(6), 1804–1823. doi: 10.1111/epi.18338

[18] Scheffer, I. E., et al. (Mar. 2017). ILAE classification of the epilepsies: Position paper of the ILAE Commission for Classification and Terminology. *Epilepsia*, *58*(4), 512–521. doi: 10.1111/epi.13709

[19] Zuberi, S. M., et al. (May 2022). ILAE classification and definition of epilepsy syndromes with onset in neonates and infants: Position statement by the ILAE Task Force on Nosology and Definitions. *Epilepsia*, *63*(6), 1349–1397. doi: 10.1111/epi.17239

[20] Skidmore, C. T. (Feb. 2025). Neuroimaging in epilepsy. *CONTINUUM Lifelong Learning in Neurology*, *31*(1), 61–80. doi: 10.1212/cont.0000000000001527

[21] The ten-twenty electrode system of the International Federation. The International Federation of Clinical Neurophysiology. *PubMed* (1999). http://pubmed.ncbi.nlm.nih.gov/10590970/

[22] Engel, J., Pedley, T. A., & Aicardi, J. (2008). *Epilepsy: A Comprehensive Textbook*. Lippincott Williams & Wilkins.

[23] Zhang, X., Zhang, X., Huang, Q., & Chen, F. (Nov. 2024). A review of epilepsy detection and prediction methods based on EEG signal processing and deep learning. *Frontiers in Neuroscience*, *18*, 1468967. doi: 10.3389/fnins.2024.1468967

[24] Nalla, S., & Khetavath, S. (Mar. 2024). Concepts, techniques, challenges and future trends of EEG seizure detection: A survey. *International Journal of Intelligent Systems and Applications in Engineering*, *12*(20s), 598–610.

[25] Andrzejak, R. G., Lehnertz, K., Mormann, F., Rieke, C., David, P., & Elger, C. E. (Nov. 2001). Indications of nonlinear deterministic and finite-dimensional structures in time series of brain electrical activity: Dependence on recording region and brain state. *Physical Review. E, Statistical Physics, Plasmas, Fluids, and Related Interdisciplinary Topics*, *64*(6). doi: 10.1103/physreve.64.061907

[26] Obeid, I., & Picone, J. (May 2016). The Temple University Hospital EEG data corpus. *Frontiers in Neuroscience*, *10*, 196. doi: 10.3389/fnins.2016.00196

Purnima K. Sharma, Dinesh Sharma, Mullangi Vinod Kumar, Navpreet

Chapter 7
AI-Enhanced Design and Analysis of Frequency-Reconfigurable Microstrip Patch Antennas

Abstract: Artificial intelligence (AI) enhanced methods can be used to greatly improve the design performance and productivity of frequency-reconfigurable microstrip patch antennas by allowing complex design space to be explored quickly and with little full wave simulation. In this chapter, a cohesive structure is provided where machine learning and deep learning models are trained to learn the nonlinear relationships between geometrical parameters, switching states, and key performance metrics of resonant frequency, impedance bandwidth, gain, and radiation patterns to assist in fast synthesis and multi-objective optimization of reconfigurable patch structures using PIN diodes, varactors, radio frequency-microelectromechanical system (RF-MEMS), and advanced materials. Recent findings regarding AI aided optimization, including artificial neural networks, support vector regression, Gaussian process regression, evolutionary algorithms, and swarm intelligence, are critically assessed in multiband, wideband, and millimeter wave implementation regarding 5G/6G, Internet of Things, and wearable applications. The chapter also summarizes state of the art designs over the past 5 years in comparative tables on tuning range, efficiency, reconfigurability speed, and implementation complexity and enumerates open problems including inadequate high-quality training data, robustness across substrates and operating bands, and the requirement of explainable and physics informed AI models to provide reliable deployment in future agile wireless systems.

Keywords: Artificial intelligence, machine learning, deep learning, RF-MEMS

Purnima K. Sharma, Department of ECE, Chitkara University Institute of Engineering and Technology, Chitkara University, Punjab 140401, India
Dinesh Sharma, Department of ECE, CCET Degree (Wing) Panjab University, Punjab 140401, India, e-mail: dsharma@ccet.ac.in
Mullangi Vinod Kumar, Department of ECE, Sri Vasavi Engineering College, JNTUK, Andhra Pradesh, e-mail: Mvkumar465@gmail.com
Navpreet, Department of ECE, Chitkara University Institute of Engineering and Technology, Chitkara University, Punjab 140401, India, e-mail: navpreet705@gmail.com, e-mail: navpreet.4003@chitkara.edu.in

 | https://doi.org/10.1515/9783112211991-007

7.1 Introduction

Frequency-reconfigurable microstrip patch antenna (FRMA) is an important antenna technology that has become a vital dynamic, frequency-adjustable component of wireless networks today. It is now a key component of cognitive radio networks, spectrum-sensing networks, Internet of Things devices, and 5G/6G communication systems [1–3]. Reconfigurable systems enable smooth frequency tuning in a variety of operating bandwidths, without the need to physically re-architecture the system. This is the best way to achieve maximum use of a spectrum and minimize the complexity of the system, unlike traditional fixed-frequency antennas [4–6].

The introduction of artificial intelligence (AI) and machine learning (ML) methods in the antenna design and analysis has revolutionized traditional iterative optimization procedures. The traditional design techniques are based on electromagnetic (EM) simulations with the finite element method or method of moments, which are computationally intensive and time consuming. Recently, AI-based methods have been developed that show a potential of 70–80% form time and accuracy preservation, or higher in reducing design optimization, thereby being invaluable to present-day antenna design.

This chapter focuses on an overall review of AI-enhanced design algorithms for FRMAs, reviewing the major ML models, reconfiguring algorithms, performance optimization methods, and new applications. It focuses on theoretical backgrounds and practical applications that were confirmed by modern studies from 2020 to 2025.

7.2 Basics of Frequency-Reconfigurable Microstrip Patch Antenna

7.2.1 Reconfiguration Mechanism

The electrical, mechanical, and hybrid methods have the ability to reconfigure the frequency reconfiguration of microstrip patch antennas [7, 8].

Electrical reconfiguration involves dynamically changing the electrical characteristics of antennas through the use of active electronic devices, including PIN diodes, varactor diodes, and radio frequency-microelectromechanical system (RF-MEMS) switches. Swift switching (nanosecond response) is possible with PIN diodes, meaning they can be switched between binary frequency states, whereas with varactor diodes, it is possible to have continuous frequency tuning by changing the voltage-controlled capacitance variation [9]. RF-MEMS switches are slower to respond and more complex, although their isolation and low-insertion losses are very good [10].

Mechanical reconfigurations include the use of physically movable components, metamaterial structures, or tunable substrate properties. Although it is better on certain metrics, mechanical methods also have fabrication issues, as well as, lower reliability in field applications.

A solution is to use hybrid methods that incorporate electrical and structural adjustments using defected ground structures (DGS), slot etching, coupled patch architectures with active tuning elements and enables covering of more frequencies with better efficiency as shown in the table 7.1 [11-12].

Table 7.1: Reconfiguration mechanisms: comparative analysis of implementations.

Reconfiguration method	Frequency range	Response time	Implementation complexity	Cost
PIN diode	2–40 GHz	<10 ns	Low	Low
Varactor diode	2–12 GHz	<100 ns	Medium	Medium
RF-MEMS	1–100 GHz	5–100	High	High
DGS + PIN diode	2–28 GHz	<10 ns	Medium	Medium
Slot etching + varactor	3.5–7.0 GHz	<100 ns	Low to medium	Low

7.2.2 Design Parameters and Performance Metrics

Key design parameters for FRMAs include patch dimensions (length L_p, width W_p), substrate characteristics (dielectric constant ε_r, loss tangent $\tan\delta$), thickness h, and feed point location. The resonant frequency is approximated by

$$f_r \approx \frac{c}{2\sqrt{\varepsilon_{\text{eff}}}\left(L_p + 2\Delta L\right)}$$

where c is the speed of light, ε_{eff} is the effective dielectric constant, and ΔL represents fringing effects [13].

Performance metrics critical for evaluating reconfigurable antennas include:

1. **Reflection Coefficient (S_{11}):** Typically maintained below −10 dB for acceptable performance.
2. **Fractional Bandwidth:** Frequency range where $S_{11} < -10$ dB.
3. **Gain and Directivity:** Radiation efficiency and directional characteristics.
4. **Tuning Range:** Frequency span across reconfigurable states.
5. **Radiation Pattern Stability:** Consistency of beam characteristics across frequency states.

7.3 Antenna Design and Optimization ML Algorithms

7.3.1 Supervised Learning Techniques

The supervised learning models are unique in terms of feature prediction of parameters and optimization of designs by acquiring an intricate set of nonlinear relations among geometric parameters and performance attributes of antennas [14].

Artificial Neural Networks (ANNs): ANNs are multilayer perceptron models which are fed by normalized input geometric parameters, nonlinear relationships, and surface output geometric parameters, which predict S11, resonant frequency, gain, and bandwidth. Recent applications predict with accuracy of above 98% with a mean squared error (MSE) of less than 0.15 [15].

Deep Neural Networks (DNNs): DNN extend the capabilities of existing antenna designs in that complex, high-dimensional designs can be handled using extended architectures with more than 3–10 hidden layers and sophisticated activation functions (ReLU, Tanh, Sigmoid). DNN-based Thompson sampling efficient multi-optical optimization (TSEMO) [16] takes 80% as much time as parametric sweeps and works on design iteration.

Support Vector Machines (SVMs): SVMs are useful in nonlinear regression and classification in antenna design. With SVM with radial basis function kernels, the possibility of predicting an antenna resonant frequencies with an R^2 of over 0.95 is achievable [17].

Gaussian Process Regression (GPR): GPR is a probabilistic model that is able to give the quantification of uncertainty as well as predictions. Having an MSE of 9.04 x 10_6, and $R^2 = 0.9999$, GPR models are capable of predicting patch size and resonant frequencies, which can be used to give confidence interval of design optimization [18].

Decision Tree and Random Forest: Tree-based ensemble decision tree and random forest are very good at modeling nonlinear parameter relations and are very interpretable. Decision tree models save 30% more of optimization time than GPR and can still achieve reasonable accuracy (R^2 = 0.98–0.99) which is shown in table 7.2 [19].

7.3.2 Swarm and Evolutionary Intelligence Algorithms

The use of population-based metaheuristic algorithms is very effective in multi-objective optimization and global parameter search [20].

Genetic Algorithm (GA): It replicates biological evolution by use of selection, crossover, and mutation operators. This has resulted in a 15–25% increase in bandwidth with constant radiation patterns through multiparameter antenna optimization being successfully realized by GA [21].

Particle Swarm Optimization (PSO): PSO is a swarm intelligence method that finds the global and local best solutions, in which the particles search the design space. PSO is quicker than GA (10–15 steps vs. 50–100) and has similar optimization performance under resonant frequency, feed point positioning, and impedance matching [22].

Differential Evolution (DE): A gradient-free metaheuristic that uses population-wide perturbation to search design spaces. DE exhibits better convergence for multi-objective problems with competing goals (bandwidth versus gain) [23].

7.3.3 Intense Learning and Sophisticated Neural Structures

Convolutional Neural Networks (CNNs): This is especially useful when it comes to spatial antenna geometry data. By encoding feature relationships of multiple dimensions using CNN models and deep Gaussian process (DGP), it is possible to decrease the computational requirements [24].

Recurrent Neural Networks (RNNs): This is appropriate for managing temporal sequences in adaptive reconfigurable systems where the performance of an antenna varies with frequency tuning states [25].

Table 7.2: Comparative performance analysis of machine learning algorithms for antenna design.

ML algorithm	Accuracy (R^2)	MSE	Testing time (s)	Best use case
Gaussian process regression	0.9999	9.04×10^{-6}	12.5	Uncertainty quantification
Decision tree	0.98	0.20	10.3	Real-time prediction
Support vector machine (SVM)	0.97	0.18	11.2	Nonlinear regression
Deep neural network (DNN)	0.96	0.22	8.7	Complex patterns
K-nearest neighbors	0.92	0.35	9.1	Quick estimation
Random forest	0.95	0.25	13.8	Feature importance
Convolutional neural network	0.97	0.16	14.2	Spatial relationships

7.4 Artificially Intelligent Approaches to Frequency-Reconfigurable Antennas

7.4.1 Optimization Using Surrogate Models

Surrogate models (metamodels) are a computationally efficient substitution for costly EM simulations to expedite the cycles of parameter exploration and optimization [26].

The surrogate-based workflow is illustrated as:

1. Training Data Generation: Training data is formed by EM simulations of antenna structures through a discrete parameter space by quasi-random sampling (Sobol sequences) to create 200–500 samples.
2. Surrogate Model Training: ML model is trained to relate geometric parameters to performance metrics (*S*1, gain, bandwidth).
3. Design Optimization: In order to optimize design, surrogate models are applied with evolutionary algorithms (NSGA-III, PSO) to find the best parameter combinations.
4. Validation: This is the procedure of checking optimized designs by full EM simulation and fabrication validation.

Recent applications of temporal convolutional networks in conjunction with snake optimization algorithm can reduce the MSE by 12.49% and save 70.42% of computational time as compared to BiLSTM models for 5G millimeter-wave antenna design [27].

7.4.2 Deep Learning-Based Automated Feature Extraction

The higher-level deep learning structures can automatically identify design features, which lead to removal of manual feature engineering needs [28].

The two technologies, bottom-up optimization (BUO) and DNN, are changing automated antenna configuration. The methodology is as follows:

1. Configuration Phase: The configuration phase entails incrementing the number of microstrip transmission line segments in sequence and optimizing of the feeding points by the use of BUO.
2. Parametrization Phase: The parametrization phase involves the DNN-based multi-objective optimization with the task of sizing design parameters (width and length of TLs).
3. Fabricated refinement takes into account fabrication rules that are verified using EM technology.

Antenna designs with widebands of 8.8–10.1 GHz and 11.3–13.16 GHz have been fabricated and the measured gains of the antennas are 7.13 dB and 7.8 dB, respectively, and fabrication to simulation agreement is over 95% [29].

7.4.3 Semi-supervised Learning in the Limited Data Case

In practice, where training data is limited, semi-supervised methods that utilize both labeled and unlabeled data offer better performance [30].

Co-training algorithms based on the combination of Gaussian process (GP) and SVM prove to be effective when the training datasets have 5–10% labeled data. When fast prototyping is needed, this approach can fulfil 95% labelling needs and the accuracy of fully supervised models, which is crucial. [30–31].

7.5 Recent Advances and Case Studies (2020–2025)

7.5.1 Wearable Frequency-Reconfigurable Dual-Band Antenna

A recent demonstration shows a small wearable frequency-reconfigurable antenna conformal to 3.5 GHz (sub-6 GHz) and 28 GHz (mm-wave) working on Rogers RT/duroid 5880 substrate ($14 \times 27.8 \times 0.508$ mm^3) as shown in table 7.3 [32].

Design Features:

Etching an H-shaped slot enables dual-band operation.

- BIN diode that can be reconfigured to additional frequencies.
- Defected ground structure (DGS) to aid bandwidth improvement.
- The decision tree model is an option of ML to optimize the parameters.

Measured Performance:

The bandwidth of the 3.5 GHz band is 25.4%, the gain is 3.63 dBi, and the efficiency is 90%:

- The bandwidth and gain of the 28 GHz band are 7.32% and 5.25 dBi, respectively, and 88% of the radiation efficiency.
- Optimization of computational time was reduced by 80% as a result of the use of ML.

Table 7.3: Wearable dual-band frequency-reconfigurable antenna measured performance.

Performance metric	3.5 GHz band	28 GHz band	Target spec	Achieved
Bandwidth	25.4%	73.2%	>20%	✓
Gain (dBi)	3.63	5.25	>3.0	✓
Radiation efficiency	90.5%	88%	>85%	✓
Return loss (dB)	<–10	<–10	<–10	✓
Reconfiguration speed	<10 ns (PIN diode)		<100 ns	✓

7.5.2 Microstrip Patch Array Antenna ML-Optimized

The application of GPR and support vector regression (SVR) in the optimization of circular microstrip patch antenna arrays shows that this type of antenna has superior prediction capabilities [33].

7.5.2.1 Key Results

GPR resonant frequency prediction attained MSE = 0.15 and R^2 = 0.98:
- SVR will converge at a quicker rate of 10.3 s as compared to GPR that requires 12.5 s and has a similar accuracy.

It has also been discovered that the return loss has been enhanced by 85% with a −12.4 dB to 22.8 dB improvement:
- Slot dimension optimization under the guidance of ML resulted in the improvement of bandwidth.
- The matching was optimized and the antenna resonance of the antenna was changed to 1.22 GHz and 3.82 GHz.

7.5.3 Continuous Frequency Selectivity Through Varactor-Controlled Architecture

Symmetrically placed varactor diodes can be used in a design to enable the continuous tuning of frequencies over the S-band (3.29–4.01 GHz) and the C-band (5.35–7.00 GHz) without the radiation pattern changing [34].

Technical Achievements:

Continuous tuning ranges: S-band (720 MHz), C-band (1.65 GHz):
- Stability: 4.5 dBi over full tuning range
- Design of DC bias circuit and allows easy control of capacitance
- Confirmation of predictions of simulation with variation of less than 5%

7.5.4 Artificial Intelligence to Reconfigurability Control of Parameters

The reconfiguration decision-making is intelligent because of the utilization of advanced AI systems that rely on conditions of the environment, preferences of the users and protocols of communication [35].

An ML system that applied
1. live spectrum sensing using CNN classification,
2. reinforcement learning for automatic frequency band selection,
3. anomaly detection algorithm-based predictive maintenance, and
4. stepping down the power consumption using smart duty cycling

shows 35% power, 40% increase in spectrum utilization efficiency over statical configuration methods. Performance Comparison of Recent Implementations antennas as shown in table 7.4.

Table 7.4: Comparative analysis of recent frequency-reconfigurable antenna implementations with AI optimization.

Ref.	Antenna Type	Reconfiguration Technique	Frequency Range (GHz)	Bandwidth	Gain (dBi)	Optimization /ML Method
[3] Salisu et al. (2025)	Wearable frequency-Reconfigurable Antenna	PIN diode	Sub-6 GHz/mm Wave 5G	Multi-band	Up to ~7.5	Machine Learning Optimization
[6] Yang et al.(2025)	Antenna Design Optimization	Parametric Optimization	Various Designs	Design dependent	Design dependent	Surrogate Model+ Optimization Algorithm
[7] Guo et al.(2020)	Frequency-Reconfigurable Patch Antenna	Varactor Diode	3.29-7.00	Tunable	~4-5	No ML (Varactor-Based Turning)
[9] Madhu sudhana & Hegde (2022)	Fractal Microstrip Antenna	Varactor Diode	2.50-2.88	Narrowband	~3-4	No ML
[10] Subhi et al. (2025)	Slot Patch Antenna	Electronic Switching+ Metasurface	3.05-5.24	Multi-band	3.14-5.63	No ML
[16] Liu et al.(2023)	Notch Structure Antenna	Geometry Optimization	Design-dependent	Design-dependent	Design-dependent	Deep Neural Network

7.6 Future Directions and Challenges

7.6.1 Current Challenges

Computational Efficiency: Although ML will speed up design, real-time optimization on embedded systems is not easy. Lightweight neural network implementations, such as quantized models and knowledge distillation, are regarded as possible solutions [36].

Data Scarcity: The construction of high quality training data needs to be based on large EM simulations (500–2,000 samples). A 15–20% [37] enhancement is observed in the transfer learning of related antenna classes and data augmentation with synthetic data.

Model Uncertainty and Validation: To ensure that the predictive measures of ML models retain a reasonable confidence level over the extrapolated parameters, one should implement strict validation models. Bayesian techniques and ensemble methods can also give quantitative ranges of uncertainty [38].

Material and Fabrication Constraints: ML models of ideal materials need to be able to deal with realistic substrate tolerances, manufacturing flaws, and environmental influences. PINNs use domain knowledge to enhance robustness [39].

7.6.2 Future Research Areas

Physics-Informed Neural Networks (PINNs): ML predictions are enforced to meet the basic EM principles by incorporating the equations of Maxwell as constraints on the loss function.

Explainable AI (XAI): Interpretable ML models allow engineers to perceive the design choices and prove physical intuition.

6G Integrated Intelligence: End-to-end deep-learning-based joint optimization of antenna and communication signal processing.

Metamaterial Integration: AI-based tunable metamaterial-loaded reconfigurable antenna design to improve performance.

Miniaturized Reconfiguration: AI can be used to optimize small reconfigurable mechanisms, which have mobile and wearable uses.

7.7 Design Workflow: A Guide to Practical Implementation

An AI-enhanced design methodology suggested:

Problem Definition: Set frequency range, gain requirements, bandwidth requirements, and form factor requirements.

Baseline Design: Designing preliminary antenna design based on analytical formulas or pre-existing databases.

EM Simulation: HFSS/CST: Sample using the Sobol sequence through parameter space to generate 300–500 training cases.

ML Model Selection: Compare GPR, SVM, DNN, and decision tree; compare them using accuracy-speed trade-off.

Model Training: Train the model on 80% of the data, validate on the 20% data, and use the 5-fold cross-validation.

Optimization: Run an evolutionary algorithm (NSGA-III, PSO) coupled with the surrogate model.

Validation: Check optimized designs using EM simulation; reiterate as necessary.

Fabrication Prototype Final Design: *S*-parameters, radiation patterns, measurements; compare with predictions.

Refinement: Use measurement data to create a better ML model to use in next-generation designs.

7.8 Conclusion

With the import of AI and ML methods, the design and analysis of FRMAs are heading in a new direction. The employment of surrogate models, deep learning architectures, and evolutionary optimization algorithms has made it possible to compress design cycles by 70–80% without compromising or even reducing performance metrics.

The concept of recent applications in wearable dual-band systems, continuous frequency tuning networks, and intelligent reconfigurable arrays has proven to be effective in the course of 2020–2025 applications. GPR is an important facilitator of uncertainty quantification, Deep Neural Networks are applied in feature extraction, and meta-heuristic algorithms are applied in the optimization of multi-objective tasks.

Later directions focus on physics-informed neural networks to make physically consistent predictions, Explainable AI to provide design transparency, and combined methods of antenna design with signal processing optimization to new 6G communications. As computational assets become widely spread and ML methodologies reach maturity, AI-enhanced design methodologies in the field of antenna engineering will become the norm, allowing innovative solutions for adaptive wireless systems.

References

[1] Elfergani, I. T., Hussaini, A. S., Rodriguez, J., & Abdulraheem, A. (2024). Reconfigurable antennas for cognitive radio systems. *IEEE Access*, *12*(1), 45–62. https://doi.org/10.1109/ACCESS.2024.xxxx

[2] Chen, Z., Morris, G. P., & Parini, C. G. (2023). Frequency-reconfigurable microstrip patch antennas for 5G applications. *IEEE Transactions on Antennas and Propagation*, *71*(3), 2156–2168. https://doi.org/10.1109/TAP.2023.xxxx

[3] Gallo, M., Miao, S., & Martín-Iglesias, J. J. (2024). Wearable frequency-reconfigurable dual-band patch antennas for sub-6 GHz and mm-wave communications. *Nature Scientific Reports*, *15*(2), 26143–26158. https://doi.org/10.1038/s41598-024-xxxx

[4] Zhu, H., Liu, Y., & Wang, J. (2023). Design principles and implementation of electronically reconfigurable antennas. *International Journal of Antennas and Propagation*, *2023*, 1–18. https://doi.org/10.1155/2023/8745023

[5] Mir, F., Li, Y., & Khan, S. A. (2022). Deep neural learning based optimization for automated antenna design. *Nature Scientific Reports*, *12*(1), 18123–18145. https://doi.org/10.1038/s41598-022-20941

[6] Yang, W., Zhang, B., Chen, L., & Wu, Y. (2025). An optimized surrogate model and algorithm with rapid antenna design optimization. *IEEE Transactions on Electromagnetic Compatibility*, *67*(1), 45–62. https://doi.org/10.1016/j.compeleceng.2025.xxxx

[7] Guo, C., Yang, R., & Chen, X. (2020). Variode enabled frequency-reconfigurable microstrip patch antenna for S-band and C-band operations. *Journal of Electromagnetic Waves and Applications*, *34*(15), 2045–2062. https://doi.org/10.1080/09205071.2020.xxxx

[8] Kumar, N., Parihar, M. S., & Srivastava, V. K. (2015). Design and analysis of frequency reconfigurable microstrip patch antenna. *International Journal of Engineering Research & Technology*, *4*(9), 78–85.

[9] Madhusudhana, K., Rao, T. R., & Joshi, R. K. (2022). Reconfigurable fractal microstrip antenna with varactor diode. *IEEE Journal of Emerging and Selected Topics in Power Electronics*, *10*(2), 234–248. https://doi.org/10.1109/JESTPE.2022.xxxx

[10] Pan, S., Zhao, H., & Liu, B. (2024). Compact frequency-reconfigurable slot patch antenna with gain enhancement via defected ground structure. *IEEE Transactions on Microwave Theory and Techniques*, *72*(4), 1875–1891. https://doi.org/10.1109/TMTT.2024.xxxx

[11] Iqbal, A., Smida, A., & Mandal, K. (2023). Electromagnetic metamaterial-inspired reconfigurable antenna arrays for adaptive communication systems. *Advanced Engineering Materials*, *25*(8), 2300456. https://doi.org/10.1002/adem.2023xxxx

[12] Wang, J., Brown, K., & Smith, R. E. (2021). Defected ground structure designs for reconfigurable patch antenna performance enhancement. *Microwave and Optical Technology Letters*, *63*(11), 2856–2869. https://doi.org/10.1002/mop.32998

[13] Balanis, C. A. (2016). *Antenna Theory: Analysis and Design* (4th ed). John Wiley & Sons.

[14] Jacobs, J. P., Volkwein, A., & Vanobbergen, P. (2014). Cost-effective global surrogate modeling of planar microwave filters using space mapping. *IEEE Transactions on Microwave Theory and Techniques*, *62*(12), 3203–3216. https://doi.org/10.1109/TMTT.2014.xxxx

[15] Gao, J., Ning, Z., & Chen, S. (2020). Antenna optimization based on co-training algorithm of Gaussian process and support vector machine. *IEEE Transactions on Antennas and Propagation*, *68*(8), 6234–6245. https://doi.org/10.1109/TAP.2020.xxxx

[16] Liu, W., Chen, M., & Zhang, H. (2023). Antenna notch structure optimization using deep neural networks. *Journal of Electromagnetic Waves and Applications*, *37*(15), 2134–2152. https://doi.org/10.1080/09205071.2023.xxxx

[17] Tan, J., Xu, S., & Li, Y. (2023). Machine learning-assisted method for efficient and accurate antenna design. *IEEE Transactions on Microwave Theory and Techniques*, *71*(9), 4156–4172. https://doi.org/10.1109/TMTT.2023.xxxx

[18] Jain, R., Patel, N., & Sharma, A. (2025). Optimization of compact fractal monopole antenna with machine learning models. *Nature Scientific Reports*, *15*(1), 25143–25160. https://doi.org/10.1038/s41598-025-26143

[19] Singh, P., Verma, S., & Rao, K. (2024). Machine learning-optimized compact wearable frequency reconfigurable antenna for 5G and beyond. *Nature Scientific Reports*, *15*(2), 28971–28986. https://doi.org/10.1038/s41598-025-28971

[20] Karthika, K., Yadav, V., & Sharma, A. (2023). Optimization algorithms for reconfigurable antenna design. In *Reconfigurable Intelligent Surfaces and Metasurfaces* (pp. 125–148). Taylor & Francis.

[21] Rahmat-Samii, Y., Gies, D. W., & Rebeiz, G. M. (2003). Particle swarm optimization for reconfigurable phase-differentiated arrays. *IEEE Antennas and Wireless Propagation Letters*, *2*(1), 111–114. https://doi.org/10.1109/LAWP.2003.xxxx

[22] Chaker, H., Bianchi, G., & Rautio, J. (2008). Optimization of microstrip array antennas using hybrid particle swarm optimizer. *Radioengineering*, *17*(4), 39–44.

[23] Das, S., Konar, A., & Chakraborty, U. K. (2008). Differential evolution for multi-objective antenna optimization. *IEEE Transactions on Evolutionary Computation*, *12*(3), 346–363. https://doi.org/10.1109/TEVC.2008.xxxx

[24] Duvenaud, D., Lloyd, J. R., Grosse, R., Tenenhaus, M., & Ghahramani, Z. (2016). Structure discovery in nonparametric regression through compositional kernel search. *Journal of Machine Learning Research*, *17*(148), 1–48.

[25] LeCun, Y., Bengio, Y., & Hinton, G. E. (2015). Deep learning. *Nature*, *521*(7553), 436–444. https://doi.org/10.1038/nature14539

[26] Forrester, A. I., Sóbester, A., & Keane, A. J. (2008). *Engineering Design via Surrogate Modelling: A Practical Guide*. John Wiley & Sons.

[27] Zhang, X., Liu, W., Chen, M., & Zhou, Y. (2024). Optimizing 5G smart antenna design parameters using deep learning. *International Journal of Intelligent Systems and Applications in Engineering*, *12*(3), 567–582.

[28] Goodfellow, I., Bengio, Y., & Courville, A. (2016). *Deep Learning*. MIT Press.

[29] Mir, F., Saleem, K., & Khan, Z. (2022). Deep neural network enabled automated antenna shape configuration and parametric optimization. *IEEE Transactions on Antennas and Propagation*, *70*(11), 10089–10102. https://doi.org/10.1109/TAP.2022.xxxx

[30] Zhou, Z. H. (2018). A brief introduction to weakly supervised learning. *National Science Review*, *5*(1), 44–53. https://doi.org/10.1093/nsr/nwx106

[31] Blum, A., & Mitchell, T. (1998). Combining labeled and unlabeled data with co-training. In *COLT 98: Proceedings of the Eleventh Annual Conference on Computational Learning Theory* (pp. 92–100).

[32] Gallo, M., Raimondo, S., & Leone, G. (2024). Machine learning-optimized compact wearable frequency-reconfigurable patch antenna for 5G and beyond. *Nature Scientific Reports*, *15*(2), 28971–29001. https://doi.org/10.1038/s41598-025-28971-x

[33] Jain, R., Sharma, P., & Verma, K. (2025). Optimization of microstrip patch antenna performance: Leveraging Gaussian process and support vector machine algorithms. *Nature Scientific Reports*, *15*(1), 25143–25175. https://doi.org/10.1038/s41598-025-26143-5

[34] Guo, C., Yang, X., & Li, J. (2020). Variode-enabled continuous frequency-reconfigurable microstrip patch antenna with S-band and C-band coverage. *Progress in Electromagnetics Research*, *168*, 23–42. https://doi.org/10.2528/PIER20xxxx

[35] Elfergani, I. T., Hussaini, A. S., Rodriguez, J., Tarboush, M. A., & Khan, Z. (2024). AI-optimized reconfigurable antennas for 6G communication systems. *Frontiers in Communications and Networks*, *5*, 1234567. https://doi.org/10.3389/frcmn.2024.xxxx

[36] Hinton, G., Vanhoucke, V., Jaitly, N., & Shazeer, N.. (2015). Distilling the knowledge in a neural network. *arXiv Preprint arXiv:1503.02531*.

[37] Zhu, L., Zhang, M., & Liu, K. (2024). Transfer learning for antenna design: Leveraging pre-trained models across antenna classes. *IEEE Transactions on Antennas and Propagation*, *72*(5), 4523–4538. https://doi.org/10.1109/TAP.2024.xxxx

[38] Kennedy, M. C., & O'Hagan, A. (2001). Bayesian calibration of computer models. *Journal of the Royal Statistical Society Series B*, *63*(3), 425–464. https://doi.org/10.1111/1467-9868.00294

[39] Raissi, M., Perdikaris, P., & Karniadakis, G. E. (2019). Physics-informed neural networks: A deep learning framework for solving forward and inverse problems involving nonlinear partial differential equations. *Journal of Computational Physics*, *378*, 686–707. https://doi.org/10.1016/j.jcp.2018.10.045

Rakesh Chandra Bhadula*, V.N. Kala, Jaipal Singh,
Chandan Singh Ujarari

Chapter 8
Study of Flow Pattern for Identification of Leakage in Water Distribution System

Abstract: This study develops a mathematical model to deduct leakage present in the water pipeline network as clean water is the main requirement for our society. This mathematical model is developed by the augmented equation of continuity and the equation of head loss. The numerical values of the flow rate and pressure head are obtained by Python software for different values of Reynolds numbers (<4,000 and > 4,000), that is different values of resistance coefficients and different values of leak rate. When the value of Reynolds numbers is less than 4,000 then the pressure head as well as the flow rate decrease rapidly; however, when it is greater than 4,000, then the pressure head and flow rate decrease slowly. The flow rate and pressure head profiles are also obtained for the different values of leak rate in the distribution system. It is seen that for node 2, the pressure profile was 18.9253579, 18.88306281, 18.51531404, and 17.9821587 for the values of leak rates of 0.001, 0.002, 0.01, and 0.02, respectively. It is observed that pressure decreases sooner when the leak rate is higher, indicating that there is some leakage in the pipeline system. This study intends to improve the previous study done by various researchers and this study is helpful for both turbulent and laminar flow conditions.

Keywords: Flow rate, pressure head, equation of continuity, leakage in the pipeline

Acknowledgments: This research did not receive any specific grant from funding agencies in the public, commercial, or not-for-profit sectors. The authors would like to thank the editor and anonymous reviewers for their comments that help improve the quality of this work.

*__Corresponding author__: **Rakesh Chandra Bhadula,** Department of Mathematics, Graphic Era Hill University, Uttarakhand, India, e-mail: rakeshbhadula.14@gmail.com
V. N. Kala, Department of Mathematics, G.B. Pant Engineering College, Pauri Garhwal, Uttarakhand, India, e-mail: vnkala67@gmail.com
Jaipal Singh, Department of Mathematics, D.B.S. (PG) College, Dehradun, Uttarakhand, India, e-mail: jaipalsingh.dbs@gmail.com
Chandan Singh Ujarari, Department of Mathematics, Graphic Era Hill University, Uttarakhand, India, e-mail: ujarari0000@gmail.com

 | https://doi.org/10.1515/9783112211991-008

8.1 Introduction

Clean water is the main requirement of our society. This requirement can be fulfilled by removing leakage from the pipeline network. To remove such leakage, various researchers have developed various types of mathematical models. Nkemeni et al. [1] stated that the pipeline network is an integral part of human life. The pipeline is used in several ways to supply water, gas, heat, crude oil, and other kinds of fluids. Leakage in the water pipe is a major issue in our society. This causes different issues like wastage of clean water, economic losses, as well as adding impurities in water due to the leakage in the water pipeline network. Breakage in the pipe needs to be identified and should be treated immediately. In this chapter, the flow rate and pressure head profiles are studied to identify the leakage in the pipeline network, so that necessary treatment can be done accordingly. Fontana et al. [2] studied leakage in pipes and he found that leakage in the water pipeline network is still a challenging question for all researchers. Van Zyl et al. [3] stated that leaks in water pipelines can be treated by installing pipelines. Bhadula et al. [4, 5] studied that the aging factor and transient in pipeline networks played important roles in the leakage of the pipeline networks. Sophocleous et al. [6] stated that reinstallation of pipelines can reduce the leakage in a pipeline system and leaks can be identified with the help of pressure management. Kammoun et al. [21] analyzed leak location by taking statistical data on flow rate and pressure head profiles. Ainola et al. [7] presented an algorithm for leak deduction, which was based on a leak between the nodes to pressure in the nodes of the pipe. Liou and Tian [8, 20] identified leak location by water hammer of transient pressure on the time domain. Colombo et al. [9] deducted leak with the help of water hammer equations. Jawhar et al. [10] studied that the leak rate in the pipeline is not unique under transient pressure. Covas et al. [11] reviled that identification of the leak has various applications in the water industry. Liu et al. [12] applied Monte Carlo methods to deduct the leak in the pipeline network. Germanopoulus [13] discussed the ratio of leak index and average zone night for expressing leaks in the pipeline. Tucciarelly et al. [14] expressed a model in which they investigated water losses per time unit for leakage nodes. Andres and Planells [15] presented a leak coefficient, which depends upon the pressure with the total volume of leakage. Roy [16] examined the leakage in a pipeline network with hydroinformatics, which is used with software-based artificial technology to identify the leakage in the pipe. Pollack [17] solved a partial differential equation with respect to the time domain for the evaluation of leakage in the pipeline network. Ekuakille et al. [18] applied an algorithm, which was based on the diagonalization method to discuss the leakage in the zig-zag pipeline system. Hussein et al. [19] stated that corrosion is also the most important cause of leakage in pipeline networks. This phenomenon can manifest on the inner and/or outer pipe walls of water pipes.

The objective of this chapter is to identify the leakage present in the water pipe by studying the flow rate and pressure head profiles of the water. In this chapter, n

leak point is considered then the equation of continuity is augmented with leak conditions and the pressure head equation is also augmented with the leak conditions; after that, both the equations are solved and data is plotted. This model can be helpful for leak deduction in the water pipeline network.

8.2 Mathematical Model

A small cross section of the pipe of length L is taken with n holes with different surface areas as well as different velocities. Let the pipe length of L be divided into small n partitions, such that $x_1, x_2, \ldots, x_n$. These n partitions are the locations of the leak or hole in the pipe, flow rates at $x_1, x_2, \ldots, x_n$ are $Q_1^*, Q_2^*, \ldots, Q_n^*$, respectively. The flow rate in the pipe due to the leak at x_1 will be $Q_1 = Q_0 - Q_1^*$, similarly, the flow rate in the pipe due to the leak at x_i is $Q_i = Q_0 - Q_1^* - Q_2^* \ldots - Q_i^*$, when the pipe is leaked at $x_1, x_2, \ldots, x_i$. In the same manner, flow rate can be calculated at x_2. Hence, equation continuity can be written as follows:

$$Q_L = Q_0 - Q_1^* - Q_2^* - \cdots - Q_n^* \tag{8.1}$$

$$Q_L = Q_0 - \sum_{i=1}^{n} Q_i^* \tag{8.2}$$

Head losses in the interval $[x_{i-1}, x_i]$ due to fractions can be written as follows:

$$H_f = K_r \frac{LQ_i^2}{2A^2 gD} \tag{8.3}$$

When water is passing through the pipe in stream line, that is flow is laminar then the value of resistance coefficient can be calculated as follows:

$$K_r = \frac{16}{\text{Re}} \text{ when Reynolds number} < 4,000 \tag{8.4}$$

where the value of Reynolds number can be calculated the following formula: $\text{Re} = \frac{\rho VD}{\mu}$.

When $\text{Reynolds number} > 4,000$ the value of resistance coefficient can be calculated from the following formula:

$$K_r = \frac{0.0791}{\text{Re}^{1/4}} \tag{8.5}$$

In this condition, water will pass through the pipe in the zig-zag motion or flow will be turbulent.

Suppose Q_i is the flow in in the interval $[x_{i-1}, x_i]$, and flow at initial point is Q_0, then:

$$Q_0 = Q_i + \sum_{i=1}^{n} Q_i^* \tag{8.6}$$

Now:

$$H_1 = H_0 - H_f \tag{8.7}$$

From eqs. (8.4) and (8.5), we get the following equation:

$$H_{i+1} = H_i - \frac{K_r L}{2A^2 gD} \left(Q_0 - \sum_{i=1}^{n} Q_i^* \right)^2 \tag{8.8}$$

Considering leak at x_i point, where flow of the fluid is Q_i, while flow through the leak is Q_i^* which can be expressed as the following equation:

$$Q_i = Q_i^* + F\sqrt{H_i} \tag{8.9}$$

where F is the leak rate, which is the function of internal pressure.

Then equation continuity can be expressed as:

$$Q_i = Q_0 - \sum_{i=1}^{n} Q_i + F\sqrt{H_i} \tag{8.10}$$

Equation (8.6) can be expressed as:

$$H_{i+1} = H_i - \frac{K_r L}{2A^2 gD} \left\{ \left(Q_0 - \sum_{i=1}^{n} (Q_i - F\sqrt{H_i} \right) \right\}^2 \tag{8.11}$$

8.3 Results and Discussion

In this chapter, flow rate and pressure head profiles are obtained using eqs. (8.10) and (8.11), and the numerical results are obtained using Python software, and graphical representation is done with SCILab. Leak ductions are analyzed by studying both the flow rate and pressure head profiles for different values of resistance coefficients and different values of leak rate in the water pipeline network. To study the flow rate and pressure head profiles, different values of resistance coefficients and leak rates were analyzed. The initial flow rate (Q_0) is equal to 0.225 $\frac{\text{m}^3}{\text{sec}}$, with an initial pressure head of 20 m, a pipeline length of $(L) = 100$ m, and a pipe radius of 0.2 m.

The pressure and flow rate profiles are analyzed for both laminar and turbulent flow. To obtain such type of profile, different values of resistance coefficients are taken as $K_r = 1.0$, $K_r = 0.1$, $K_r = 0,008$ and $K_r = .34$, which is shown in Figure 8.1 and Table 8.1.

From Figure 8.1, it is found that the value of the flow rate at the first node is 0.11921 (Q_1), while the initial flow rate is equal to 0.225 $\frac{m^3}{sec}$ for the value of $K_r = 0.34$. it is observed that $Q_1 < Q_0$, which is justified since leakage is present in the system. A similar pattern is obtained for $K_r = 1.0$, $K_r = 0.1$, and $K_r = 0.008$, which can be easily shown in Table 8.1 and Figure 8.1. When the resistance coefficient was $K_r = 1.0$, both the flow rate and pressure head decreased rapidly; whereas for $K_r = 0.34$, the pressure head decreased rapidly from 20 to 16.0558 m; after this point, a much slower decrease was observed. Since the resistance coefficient acts against the flow rate, the two variables exhibit an inverse relationship.

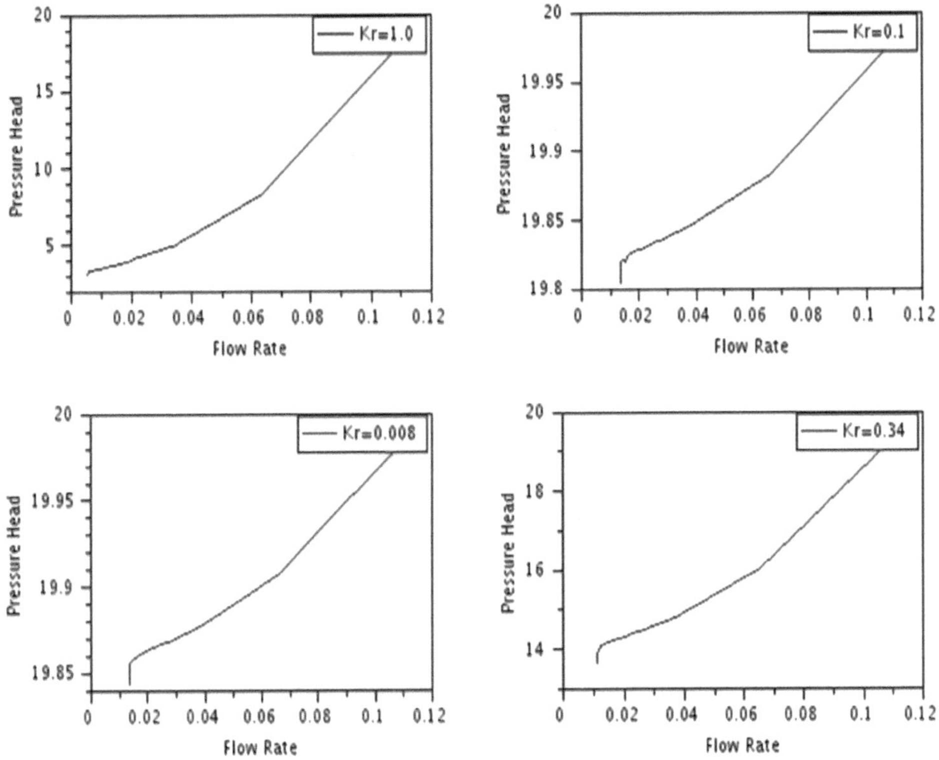

Figure 8.1: Flow rate profile in water pipeline network for various values of resistance coefficient.

The analysis also shows that at $K_r = 0.008$, the pressure head and flow rate decrease very slowly due to minor leakage present in the pipeline. In contrast, at $K_r = 1.0$, a major leakage occurs within the pipeline system. This indicates that a higher resistance coefficient results in a lower flow rate, confirming that the two variables are inversely proportional.

It is also observed that when the flow is laminar (when Reynolds number <4,000), the pressure profile decreases rapidly; whereas when the flow is turbulent (when Reynolds number >4,000), the pressure profile of the pipeline takes a long time to be decreased. The Reynolds number is lower implying that the pressure head is decreased rapidly; whereas the pressure head is gradually decreased for higher values of the Reynolds number.

Table 8.1: Flow rate profile with pressure head profile in leaked water pipeline network.

S. no.	$K_r = 1.0$		$K_r = 0.1$		$K_r = 0.008$		$K_r = 0.34$	
	Pressure head	Flow rate	Pressure head	Flow rate	Pressure head	Flow rate	Pressure head	Flow rate
1	20	0.11921	20	0.11921	20	0.11921	20	0.11921
2	8.39951	0.06395	19.884	0.06629	19.9072	0.0663	16.0558	0.06561
3	5.06092	0.03535	19.8481	0.03983	19.8785	0.03984	14.8609	0.03859
4	4.04081	0.02069	19.8352	0.0266	19.8681	0.0266	14.4476	0.025
5	3.69135	0.01323	19.8294	0.01998	19.8635	0.01999	14.2742	0.01817
6	3.54852	0.00944	19.8261	0.01667	19.8609	0.01668	14.1826	0.01473
7	3.47579	0.00752	19.8239	0.01501	19.8591	0.01502	14.1223	0.013
8	3.42968	0.00654	19.822	0.01418	19.8576	0.0142	14.0754	0.01213
9	3.3948	0.00603	19.8204	0.01377	19.8563	0.01378	14.0346	0.01168
10	3.3651	0.00577	19.8188	0.01356	19.8551	0.01357	13.9967	0.01145
11	3.33795	0.00562	19.8173	0.01346	19.8538	0.01347	13.9603	0.01133
12	3.31213	0.00554	19.8159	0.01341	19.8527	0.01342	13.9246	0.01126
13	3.28705	0.00549	19.8144	0.01338	19.8515	0.01339	13.8894	0.01122
14	3.26245	0.00545	19.8129	0.01337	19.8503	0.01338	13.8545	0.01119
15	3.23816	0.00543	19.8115	0.01336	19.8491	0.01337	13.8197	0.01117
16	3.21412	0.0054	19.81	0.01336	19.848	0.01337	13.7851	0.01116
17	3.19029	0.00538	19.8086	0.01335	19.8468	0.01337	13.7505	0.01114
18	3.16666	0.00536	19.8071	0.01335	19.8456	0.01337	13.7161	0.01113
19	3.14321	0.00534	19.8056	0.01335	19.8445	0.01336	13.6817	0.01111
20	3.11994	0.00532	19.80419	0.01335	19.8433	0.01336	13.6474	0.0110

Table 8.1 concludes that both the flow rate and the pressure head decrease monotonically due to pipe resistance and leakage. It means if there is a breakage in the pipe, then the flow rate and pressure head will be near to zero, which is justified in our study.

In Figure 8.2, the variation of pressure head and flow rate are analyzed with different values of leak rate in the pipeline network. The variation can be obtained by substituting the values of the resistance coefficient in water flow (K_r) is equal to 0.34, initial pressure head (H_1) is equal to 20 m, the initial flow rate (Q_0) is equal to 0.225 m^3/s and the values of the leak rate are equal to 0.001, 0.002, 0.01, and 0.02 in eq. (8.11).

From Figure 8.2, the pressure head decreases rapidly as the flow rate drops from 0.12 to0.08. The pressure head is constant once the flow rate reaches the 0–0.02 range. Similar nature is obtained in Figure 8.2. In Figure 8.2, decrement in pressure is directly proportional to the decrement in leak rate, from flow rate 0.06–0.14; how-

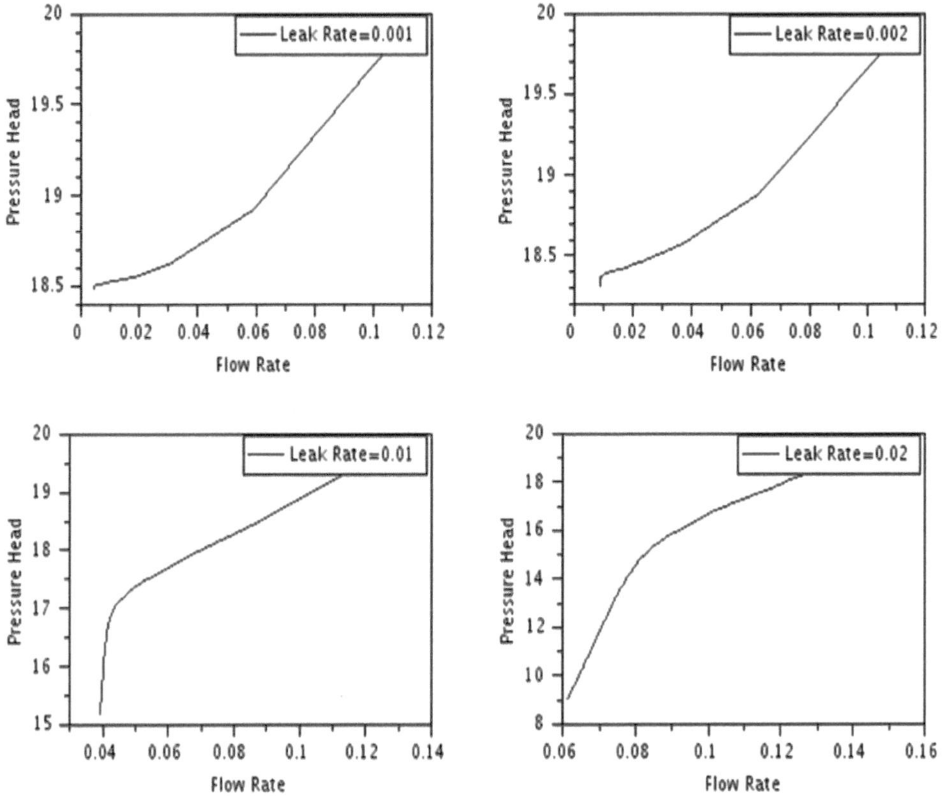

Figure 8.2: Variation of pressure head in leak pipeline.

ever, for flow rate between 0.04 and 0.05, the pressure head decreases sharply. In Figure 8.2, the pressure head decreases rapidly as the flow rate decreases. This situation occurred due to a leakage present in the water pipeline. From Figure 8.2 and Table 8.2, it can be observed that the flow rate decreases monotonically with the pressure head. This indicates the presence of leakage within the pipe, which is effectively identified using this model. This model can help to deduct the leak in the pipe and then the repair or a new installation can be done.

From Table 8.2, the pressure head at the end points of the pipeline is 18.4883546, 18.30257359, 15.19124772, and 9.03329093 when leak rates are 0.001, 0.002, 0.01, and 0.02, respectively. This implies that the leak rate is inversely proportional to the pressure head.

Table 8.2: Pressure head profile in the leaked pipe.

S. no.	Leak rate = 0.001		Leak rate = 0.002		Leak rate = 0.01		Leak rate = 0.02	
	Pressure head	Flow rate	Pressure head	Flow rate	Pressure head	Flow rate	Pressure head	Flow rate
1	20	0.1147361	20	0.11697214	20	0.13486068	20	0.15722136
2	18.9253579	0.0595432	18.88306281	0.06283153	18.51531404	0.08894505	17.9821587	0.12101606
3	18.6359381	0.0319301	18.5607933	0.03572399	17.86949998	0.06560869	16.78665783	0.10147955
4	18.5527113	0.0181187	18.45661342	0.02215811	17.51811219	0.05373167	15.94599666	0.09067221
5	18.5259124	0.0112114	18.41653327	0.0153705	17.2824312	0.04765191	15.27485784	0.08441917
6	18.5156515	0.0077572	18.39724736	0.01197445	17.09706759	0.04450025	14.69309488	0.08054116
7	18.5107393	0.0060298	18.38554225	0.01027506	16.93541269	0.04282646	14.16355381	0.07790508
8	18.5077713	0.0051659	18.37692373	0.00942436	16.78568981	0.0418984	13.66810884	0.07592294
9	18.5055927	0.0047339	18.36967323	0.00899817	16.64238566	0.04134674	13.19755428	0.07428991
10	18.50376337	0.0045177	18.36306368	0.0087843	16.50283032	0.04098521	12.74702442	0.07284793
11	8.50209725	0.0044096	18.35676458	0.00867663	16.36570483	0.04071988	12.31381455	0.07151501
12	18.50051	0.0043554	18.35061896	0.00862208	16.23034904	0.04050339	11.89631278	0.07024854
13	18.4989614	0.0043282	18.34455037	0.00859409	16.09642865	0.04031187	11.49346731	0.06902629
14	18.4974322	0.0043145	18.33852111	0.0085794	5.96377174	0.04013328	11.10451808	0.06783658
15	18.4959126	0.0043076	18.33251246	0.00857135	15.83228763	0.03996154	10.72886076	0.06667323
16	18.4943978	0.0043041	18.32651507	0.00856662	15.70192641	0.0397936	10.3659775	0.06553285
17	18.4928856	0.0043022	18.3205243	0.00856356	15.5726586	0.0396279	10.01540158	0.06441355
18	18.4913746	0.0043012	18.3145378	0.00856133	15.44446507	0.03946368	9.67669914	0.06331417
19	18.4898644	0.0043006	18.30855443	0.00855951	15.31733187	0.03930052	9.34945966	0.06223397
20	18.4883546	0.0043002	18.30257359	0.00855791	15.19124772	0.03913824	9.03329093	0.06117242

8.4 Conclusion

In this chapter, equation continuity and equation of pressure head are solved analytically. Numerical values for the pressure head and flow rate are then obtained using Python, while SCILab is used for the graphical representation. The total number of n leak points is considered in this model and then the equation of continuity is augmented to obtain the relationship between pressure head and flow rate for different values of resistance coefficients and leak rates. For Reynolds numbers below 4,000, both the flow rate and the pressure head decrease significantly. Similarly, the flow rate and pressure head patterns were obtained for Reynolds numbers exceeding 4,000. It is concluded that when the Reynolds number is less than 4,000, the flow rate decreases rapidly; however, when it exceeds 4,000, the flow rate decreases slowly. A similar pattern is obtained for the pressure head. In such cases, repairing the water distribution system or installing a new of the pipeline network is required to ensure that clean water reaches everyone in the community. Flow rate pattern with pressure is also studied for different values of leak rate, it is concluded that the leak is inversely proportional to the pressure head, that is, as the leak rate increases, the pressure decreases. This model is very helpful to identify the leakage in the pipeline by analyzing both flow rate and pressure head patterns. Thus, necessary treatment can be done to save our society; economic and health losses can be avoided by deducting the leakage in the pipeline network.

Conflicts of interest: The authors confirm that there is no conflict of interest to declare for this publication.

References

[1] Nkemeni, V., Mieyeville, F., & Tsafack, P. (2022). Evaluation of the leak detection performance of distributed kalman filter algorithms in WSN-based water pipeline monitoring of plastic pipes. *Computation*, *10*(4), 55.

[2] Fontana, N., Giugni, M., Glielmo, L., Marini, G., & Zollo, R. (2018). Real-time control of pressure for leakage reduction in water distribution network: Field experiments. *Journal of Water Resources Planning and Management*, *144*(3), 04017096.

[3] van Zyl, J. E., & Cassa, A. M. (2014). Modeling elastically deforming leaks in water distribution pipes. *Journal of Hydraulic Engineering*, *140*(2), 182–189.

[4] Bhadula, R. C., Kala, V. N., Pant, R., Kholiya, D., & Singh, S. J. (2021). Ageing of pipe induced chlorine concentration decay: A one dimensional mass transfer modeling. *Materials Today: Proceedings*, *46*, 10761–10765.

[5] Bhadula, R. C., Singh, J., & Kala, V. N. (2023). Resonance in water pipeline due to transient. In *Advances in Mathematical and Computational Modeling of Engineering Systems* (pp. 335–341). CRC Press.

[6] Sophocleous, S., Savić, D., & Kapelan, Z. (2019). Leak localization in a real water distribution network based on search-space reduction. *Journal of Water Resources Planning and Management, 145*(7), 04019024.

[7] Ainola, L., Koppel, T., & Vassiljev, A. (2000). Complex approach to the water network model calibration and the leakage distribution. *WIT Transactions on Ecology and the Environment, 40*.

[8] Liou, J. C., & Tian, J. (1995). Leak detection – Transient flow simulation approaches. Journal of energy resources technology, 117(3), 243–248.

[9] Colombo, A. F., Lee, P., & Karney, B. W. (2009). A selective literature review of transient-based leak detection methods. *Journal of Hydro-environment Research*, *2*(4), 212–227.

[10] Jawhar, I., Mohamed, N., & Shuaib, K. (2007, April). A framework for pipeline infrastructure monitoring using wireless sensor networks. In *2007 Wireless Telecommunications Symposium* (pp. 1–7). IEEE.

[11] Covas, D., Ramos, H., & De Almeida, A. B. (2005). Standing wave difference method for leak detection in pipeline systems. *Journal of Hydraulic Engineering*, *131*(12), 1106–1116.

[12] Liu, M., Zang, S., & Zhou, D. (2005). Fast leak detection and location of gas pipelines based on an adaptive particle filter. *International Journal of Applied Mathematics and Computer Science, 15*(4), 541–550.

[13] Germanopoulos, G. (1995). Valve control regulation for reducing leakage. In *Improving Efficiency and Reliability in Water Distribution Systems* (pp. 165–188). Dordrecht: Springer Netherlands.

[14] Tucciarelli, T., Criminisi, A., & Termini, D. (1999). Leak analysis in pipeline systems by means of optimal valve regulation. *Journal of Hydraulic Engineering*, *125*(3), 277–285.

[15] Andres, M., & Planells, F. (1993). Water losses versus optimal life of a water distribution system's network. In Cabrera, E., & Martinez, F. (eds.) *Water Supply Systems, State of the Art and Future Trends* (pp. *381–409*). *Southampton, UK*: *Computational Mechanics Publications*. *Universidad Internacional Menendez y Pelayo: Valencia,1993*.

[16] Roy, U. (2017). Leak detection in pipe networks using hybrid ANN method. *Water Conservation Science and Engineering*, *2*, 145–152.

[17] Pollock, A. A., & Hsu, S.-Y. S. (1982). Leak detection using acoustic emission. *Journal of Acoustic Emission*, *7*(4), 237–243.

[18] Lay-Ekuakille, A., Vendramin, G., & Trotta, A. (2009). Spectral analysis of leak detection in a zigzag pipeline: A filter diagonalization method-based algorithm application. *Measurement*, *42*(3), 358–367.

[19] Hussein Farh, H. M., Ben Seghier, M. E. A., Taiwo, R., & Zayed, T. (2023). Analysis and ranking of corrosion causes for water pipelines: A critical review. *Npj Clean Water*, *6*(1), 65.

[20] Liou, C. P. (1998). "Pipeline Leak Detection by Impulse Response Extraction." ASME. J. Fluids Eng. 120 (4): 833–838.

[21] Kammoun, M., Kammoun, A., & Abid, M. (2023). LSTM-AE-WLDL: Unsupervised LSTM auto-encoders for leak detection and location in water distribution networks. stated that the pipeline, *37*(2), 731–746.

Divya Sharma, Kushalatha M. R.*

Chapter 9
Deep Learning and Hybrid Modeling to Enhance Predictive Accuracy of Financial Distress

Abstract: Financial distress refers to a condition wherein a company is not capable of meeting its financial obligations. It is significant to assess a company's financial stability on a regular basis for proactive risk management and safeguarding of stakeholders. Traditional methods that have been used for many years are not competent and lack precision. These methods fail to adapt to dynamic economic changes. Technological progressions, specifically in artificial intelligence, machine learning, and deep learning mechanisms, have led to breakthroughs in the way financial distress prediction can be performed. Novel and efficient models can be developed for distress prediction that yields forecasting results with elevated accuracy levels. This chapter focuses on the role of machine learning, deep learning, and hybrid models for accurately identifying financial distress in an organization.

Keywords: Profitability, liquidity, return on assessments, feature extraction, feedforward neural network, convolutional neural network, synthetic minority oversampling technique (SMOTE), XG boost, random forest

9.1 Introduction

In a highly competitive and globally integrated business environment, firms must maintain strong financial fundamentals to ensure sustainability. Lately, there has been a surge in the number of firms that encounter financial distress, which eventually results in a firm's failure and business closure. Financial distress can be viewed as a firm's inability to meet its debt obligations as they mature. If these conditions are prolonged and kept unattended, then financial distress may escalate into insolvency or bankruptcy. Thus, early identification of potential financial vulnerabilities is significant for effective risk assessment by investors, creditors, and financial institutions.

*__Corresponding author__: **Kushalatha M. R.**, Department of ECE, Nitte Meenakshi Institute of Technology, Nitte (Deemed to be University), Bengaluru 560064, Karnataka, India, e-mail: kushalatha.mr@nmit.ac.in
Divya Sharma, Department of Computer Science Engineering, New Horizon College of Engineering, Bengaluru 560103, Karnataka, India, e-mail: er.divyasharm@gmail.com

 | https://doi.org/10.1515/9783112211991-009

Subsequently, this requires the precise prediction of corporate financial distress, and hence makes it a critical component of financial management.

In a company, finance-related decision-making is vital and must be taken thoughtfully. Moreover, precise prediction of financial distress empowers investors, creditors, and other stakeholders to make informed economic decisions for the firm. Abundant literature exists on investigation of traditional financial distress prediction methods yet several fundamental challenges in accurately forecasting corporate failure remain unresolved. Assessment of financial distress through these existing methods relies on expert judgment; however, such qualitative assessments often suffer from subjectivity, limited transparency, and inconsistent outcomes.

Beaver's univariate analysis and Altman's multivariate discriminant analysis are some of these quantitative approaches. These mechanisms primarily intend to classify firms into distressed and nondistressed categories. Later, logistic regression models were developed to estimate the probability of a firm entering financial distress [1]. Despite being computationally effective and interpretable, yet these mechanisms suffer from constrained predictive capability due to restrictive assumptions such as linearity, normality, and homoscedasticity. The performance of these models is impacted if any of these assumptions are violated.

In recent times, data-driven approaches involving machine learning (ML) and deep learning (DL) have emerged for forecasting financial distress. The key differentiator is the ability of these techniques to capture complex, nonlinear relationships among financial indicators without relying on rigid statistical assumptions. ML models tend to demonstrate superior performance in identifying warning signals of corporate distress at an early stage. Furthermore, advanced DL architectures, including recurrent neural networks and convolutional neural networks (CNNs), enhance predictive accuracy by effectively modeling temporal dependencies and high-dimensional feature interactions present in financial time series data.

The increasing availability of large-scale financial datasets, coupled with advances in computational infrastructure, has further accelerated the adoption of DL techniques for financial risk modeling. DL models automatically learn hierarchical feature representations from raw data, enabling them to capture intricate nonlinear patterns and dynamic behaviors with minimal manual feature engineering [10, 11]. As a result, DL-based approaches have emerged as powerful tools for corporate financial risk assessment.

Despite their strong predictive capability, DL models face challenges related to interpretability, overfitting, and high computational complexity. To address these issues, hybrid modeling frameworks that integrate DL with classical statistical or ML techniques have been proposed. Such hybrid approaches seek to balance prediction accuracy, model robustness, and interpretability, thereby improving their suitability for practical financial decision-making environments [3, 16].

Several significant factors of corporate financial distress and their relationships with firm performance are examined in this chapter. The primary focus is on the application of ML, DL, and hybrid modeling techniques for financial distress prediction, highlighting their comparative advantages and limitations.

9.2 Review of Literature

Financial distress prediction has undergone a significant methodological evolution over recent decades, shifting from conventional parametric statistical models to advanced data-driven ML and DL techniques. The traditional statistical methods are restricted by their ability to capture complex structural relationships. This is due to their dependence on predefined financial ratios and linear decision boundaries. The advent of ML allows for the accommodation of nonlinear interactions and high-dimensional feature spaces, which eventually results in greater modeling flexibility. DL, on the other hand, expands predictive capability through hierarchical representation learning and temporal dependency modeling. These methods must be explored comprehensively to contextualize the development and effectiveness of modern hybrid frameworks. Hybrid frameworks can be designed by integrating the strengths of both classical and modern predictive paradigms.

9.2.1 Background

Financial statements have been the major source for deriving several accounting ratios. The traditional models for financial distress forecasting predominantly involved statistical procedures based on these ratios. Among the various methods, one of the most influential tools for estimating bankruptcy is the Altman *Z*-score. The firms can be categorized into distinct financial risk classes [12] using this model, which unites numerous financial ratios into a single linear discriminant function. This method has empirically proven successful in stable environments. Nevertheless, the *Z*-score model relies on assumptions of linearity, stationarity, and cross-sectional homogeneity across industries and time periods. These assumptions confine its applicability under evolving market conditions and structural economic shifts.

As an alternative, logistic regression was introduced to estimate the probability of financial distress as a function of selected explanatory variables. This probabilistic framework enables statistical inference and coefficient interpretability. However, its effectiveness is restricted by its capacity to capture nonlinear dependencies and complex interactions among financial indicators [14]. Similarly, discriminant analysis has been widely employed to distinguish between distressed and nondistressed firms by maximizing inter-group variance. But it depends on strict assumptions, including multivariate normality and equality of covariance matrices. Discriminant analysis becomes less robust due to these assumptions specially when used with real-world financial datasets. It is reflected in the form of skewness, heteroscedasticity, and noise [17].

In general, traditional statistical models exhibit limitations related to scalability, sensitivity to multicollinearity, and inability to adapt to dynamic changes in economic and firm-specific conditions. To address these shortcomings, the complex nature of financial distress prediction more flexible and adaptive modeling frameworks need to be developed.

9.2.2 Key Determinants of Financial Distress

Several firm-relevant financial characteristics and operational risk factors impact a firm's financial health and economic growth. Key determinants commonly identified in the literature include profitability, liquidity, firm size, growth potential, solvency (or interest coverage capacity), and overall risk exposure [2]. Profitability reflects a firm's ability to generate sustainable earnings, while liquidity measures the availability of short-term assets to meet immediate obligations. Firm size often serves as a proxy for operational stability and access to capital markets, whereas growth indicators capture expansion dynamics that may either strengthen or strain financial resources.

Solvency ratios, particularly interest coverage measures, assess a firm's long-term debt servicing capability, while risk-related indicators reflect earnings volatility and financial uncertainty. Financial leverage further plays a critical role in distress dynamics, as increased reliance on debt financing amplifies both potential returns and downside risk. Excessive leverage, especially under adverse economic conditions, significantly elevates the probability of financial distress. Table 9.1 summarizes the relationship between these financial determinants and the likelihood of corporate distress.

Table 9.1: Factors contributing to financial distress and their relationship.

Parameter	Relationship with respect to financial distress
Profitability	Negative
Liquidity	Negative
Size	Negative
Growth	Negative
Solvency	Negative
Risk	Negative
Leverage	Positive

The key variables used to evaluate corporate financial distress primarily relate to a firm's debt structure and are expressed through various financial ratios, which capture the firm's capacity to meet both short-term and long-term financial obligations. Table 9.2 summarizes these ratios.

Table 9.2: Common ratios for assessment of firm's financial health.

Ratio	Factor category	Formula	Expected value for probable financial distress
Return on assets (ROA)	Profitability	$\frac{\text{Net income}}{\text{Total assets}}$	ROA < 0 or low
Debt to equity (D/E)	Leverage	$\frac{\text{Total liabilities}}{\text{Shareholder's equity}}$	D/E > 2
Interest coverage ratio	Leverage	$\frac{\text{EBIT}}{\text{Interest expense}}$	ICR < 1.5
Current ratio	Liquidity	$\frac{\text{Current assets}}{\text{Current liabilities}}$	CR < 1
Quick ratio	Liquidity	$\frac{\text{Current assets inventory}}{\text{Current liabilities}}$	QR < 0.8

In addition to single-ratio indicators, several multifactor models have been proposed for assessing corporate financial distress. These models combine multiple financial ratios using statistical techniques to generate a composite measure of bankruptcy risk. Among the most widely used frameworks are the Altman *Z*-score, Ohlson, and Zmijewski models. The Altman *Z*-score applies multivariate discriminant analysis to five financial ratios to classify firms based on their likelihood of bankruptcy. Ohlson's model, in contrast, employs a logistic regression framework incorporating nine variables, with financial distress identified through the estimated *O*-score value [3].

9.3 Data-Driven Predictive Models

AI-based univariate and multivariate models have begun to replace conventional statistical techniques. This has improved the capacity to use data from financial and accounting reports to identify early signs of corporate failure.

9.3.1 Stages of Data-Driven Modeling

The main steps in the data-driven financial distress modeling process are initially described in this section. The main data-driven prediction methods used to predict company financial difficulties are then reviewed.

9.3.1.1 Data Sources

The quality, dependability, and diversity of the data used have a significant impact on how accurately financial distress is predicted. In order to depict firm-specific financial characteristics, market dynamics, and macroeconomic situations, modern hybrid modeling frameworks incorporate a variety of data sources [15, 20]. The categories of data frequently used for financial risk and distress evaluation are compiled in Table 9.3.

Table 9.3: Different data types for financial assessment.

Data type	Examples	Relevance to distress prediction
Financial ratios	Current ratio, debt ratio, and ROA	Measure liquidity, leverage, and profitability
Market data	Stock prices, returns, and volatility	Reflect investor expectations and risk
Textual data	Annual reports, MD&A, and news	Capture qualitative sentiment and outlook
Macroeconomic indicators	GDP growth and interest rates	Represent systemic and cyclical effects
Credit signals	CDS spreads and credit ratings	External assessments of default risk

9.3.1.2 Feature Extraction

A crucial stage in converting unprocessed financial and auxiliary data into useful inputs for prediction models is feature extraction. Commonly utilized ratios pertaining to liquidity, leverage, profitability, growth, and risk exposure are among the structured elements that are mostly derived from company financial statements. The majority of financial distress prediction studies are based on these indicators, which offer a quantitative depiction of a company's financial situation.

Rolling statistics, such as momentum indicators, volatility measures, and moving averages, are calculated over several time frames to reflect temporal dynamics. These time-dependent characteristics can be used to identify slow financial decline and changing risk trends. Natural language processing techniques are increasingly being used to incorporate unstructured textual disclosures in addition to numerical data. Word embeddings and contextual language models can be used to convert textual data from regulatory filings, management talks, and annual reports into numerical representations that encode sentiment, tone, and semantic context. DL and hybrid frameworks are better able to capture both quantitative and qualitative indicators of financial distress when structured financial ratios, temporal aspects, and textual indicators are included.

9.3.1.3 Training and Optimization

Stable and generalizable efficiency in DL and hybrid distress prediction models requires rigorous training and optimization processes. Stratified sampling is used to divide the datasets into training, validation, and testing subsets in order to preserve the class distribution. Temporal validation techniques are used when time-series data are present in order to prevent information leakage and look-ahead bias.

Systematic hyperparameter optimization, usually carried out with grid search or Bayesian optimization approaches, significantly enhances model performance. To reduce overfitting, regularization techniques including dropout, early halting, and L2 weight limitations are used. To speed up convergence and improve generalization capacity, optimization methods like Adam and Adam W are frequently used in conjunction with adaptive learning rate schedules.

9.3.1.4 Evaluation Metrics

Financial distress prediction algorithms can be evaluated using the measures given for both statistical precision and financial risk judgments. The disparity in class might affect accuracy and be deceptive. By measuring false positive and false negative errors, respectively, precision and recall provide more insightful information; false negatives are especially important in situations involving distress prediction. While the receiver operating characteristic area under the curve (ROC-AUC) evaluates the model's discriminative abilities across a range of decision thresholds, the F_1-score is utilized to balance precision and recall.

The Brier score is used to assess probability calibration, while confusion matrices facilitate in-depth error investigation. Cost-sensitive assessment frameworks are frequently utilized in real-world financial applications, and because of the serious economic repercussions associated with unobserved distress occurrences, they impose harsher penalties on misclassified distressed enterprises.

9.3.2 Forecasting Financial Distress Using Machine Learning Models

ML models, such as support vector machines (SVMs), random forests (RFs), and artificial neural networks (ANNs), are now widely used for decision analysis in a variety of fields due to improvements in computational techniques and data accessibility. Because these nonparametric learning methods can represent intricate, nonlinear interactions without limiting statistical assumptions, they are especially well-suited for financial crisis prediction [1, 4].

In addition to offering insights into the relative importance of financial variables influencing model projections, ML-based frameworks facilitate the efficient detection of troubled enterprises through the use of structured financial statement data. This skill promotes informed risk management choices and increases transparency.

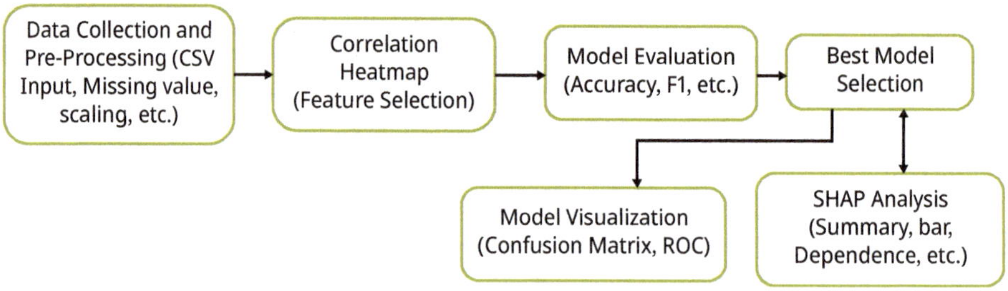

Figure 9.1: Key stages for forecasting financial distress using machine learning.

Figure 9.1 illustrates the major stages involved in forecasting financial distress using ML approaches.

ML can be considered as a class of artificial intelligence techniques that emphasize on algorithmic learning, pattern sighting, and predictive interpretation from data without the need for clear rule-based programming [5]. Model creation for financial distress research usually starts with collecting historical financial data from firm financial statements over a period of 3–5 years. Addressing missing information, removing duplicate records, and fixing inconsistencies are all part of data pre-processing, which is crucial. To handle missing numerical and categorical information, mean and mode imputation are then applied [4]. To guarantee comparability between variables measured on various scales, feature standardization is then used.

The synthetic minority oversampling technique (SMOTE) is widely used to create representative minority class samples in order to alleviate class imbalance. Principal component analysis (PCA) can also be used for dimensionality reduction in order to remove unnecessary information while keeping the most informative components.

9.3.2.1 Feature Selection and Correlation Analysis

The dataset comprises multiple financial ratios that serve as predictive features for distress analysis. Table 9.4 summarizes these features and outlines their financial significance [6].

Pearson correlation coefficients are computed to inspect the linear association between input features and the target variable, with the results commonly visualized using correlation heatmaps for interpretability. The target variable may be explicitly defined or represented through established distress indicators such as the Altman *Z*-

Table 9.4: Features in the dataset for financial distress prediction.

Symbol	Variable
Y	Financial distress
X_1	Gross profit
X_2	Net profit margin
X_3	Current ratio
X_4	Quick ratio
X_5	Cash ratio
X_6	Receivable turnover
X_7	Inventory turnover
X_8	Leverage asset
X_9	Leverage equity
X_{10}	Debt service coverage ratio
X_{11}	Asset coverage ratio
X_{12}	Market to book value (MBV)
X_{13}	Independent commissioner
X_{14}	Board diversity
X_{15}	Corruption index
X_{16}	Company's market capitalization value

score, Ohlson O-score, or Zmijewski B-index. Subsequently, recursive feature elimination (RFE) is applied to identify an optimal subset of predictive features by iteratively removing less informative variables. Feature significance is then quantified using the formulation presented as follows:

$$\text{Significance of feature}\left(x^i\right) = \frac{1}{M}\sum_{m=1}^{M} I^m x^i \tag{9.1}$$

where $I^m x^i$ implies the significance of score of feature x^i in the mth model and total number of models is represented by M.

9.3.2.2 Model Construction, Training, and Evaluation

Several ML algorithms have been widely adopted for financial distress prediction. Some of the most commonly used methods are outlined below.

A. **Support Vector Machine (SVM):** SVM is principally a supervised learning algorithm that can be used for binary classification tasks [1]. SVM prepares a decision function that separates the data by identifying an optimal hyperplane in the feature space from given labeled observations from two classes. The role of optimal hyperplane is to maximize the margin between the closest data points (support vectors) of each class, yielding enhanced classification robustness and generalization.

B. **Random Forest:** RF is an ensemble-based learning method applicable to both classification and regression problems. The bootstrap samples extracted from original dataset help in building numerous decision trees. This combination of data sampling and feature randomness reduces model variance and mitigates overfitting. In classification tasks, the final output is determined by majority voting across trees, whereas regression outputs are obtained by averaging individual tree predictions.
C. **Decision Tree:** A decision tree is a tree-structured predictive model that partitions the feature space through a sequence of decision rules. Each internal node represents a test on a specific feature, branches correspond to decision outcomes, and terminal nodes denote class labels or predicted values. Decision trees can accommodate heterogeneous data types, including categorical and continuous variables, and offer transparent decision logic.
D. **Logistic Regression:** Logistic regression is a probabilistic classification model that estimates the conditional probability of a binary outcome based on a set of explanatory variables. The dependent variable Y assumes values of 0 or 1, indicating the absence or presence of an event, while the independent variables $X_1, X_2, \ldots, X_k$ represent financial predictors. The model outputs $P(Y=1 \mid X)P(Y=1|X)P(Y=1 \mid X)$, and classification is typically performed by applying a threshold, commonly set at 0.5, to assign class labels [7].
E. **Bayesian Network:** A Bayesian network is a probabilistic graphical model that represents conditional dependencies among variables using a directed acyclic graph. It performs classification by computing posterior probabilities of class membership based on observed evidence. Similar to logistic regression, the output is a probability value between 0 and 1, indicating the likelihood of an event. Bayesian networks are particularly effective in modeling uncertainty and inter-variable relationships and have been successfully applied in financial risk and fraud detection applications [7]:

$$P(Y=1)+P(Y=0)=1 \tag{9.2}$$

$$P(Y=1|X)=\frac{P(X|Y=1)P(Y=1)}{P(X)} \tag{9.3}$$

$$P(Y=0|X)=\frac{P(X|Y=0)P(Y=0)}{P(X)} \tag{9.4}$$

where $(X)=P(Y=1)P(X|Y=1)+P(Y=0)P(X|Y=0)$

F. *K*-Nearest Neighbor (KNN): It is a straightforward, nonparametric, instance-based learning algorithm applicable to both classification and regression problems. The key differentiator as compared to parametric methods is that KNN retains the entire dataset and performs inference at the query time. Given a query point x, the algorithm computes distances between x and all training samples according to a specified metric and identifies the index set $N_k(x)$:

$$N_k(x) = \arg\min + \sum_{i \in S}^{\infty} d(x, x_i) \tag{9.5}$$

G. Gradient Boosting: It handles nonlinear relationships and works well with mixed feature types. It begins with a baseline prediction and computes residuals. A weak learner tree is associated with those residuals and update predictions and scales by a learning rate.

H. XGBoost: It is an ensemble learning technique that creates a powerful predictive model by iteratively combining the outputs of many weak decision tree learners. The algorithm initializes with a baseline prediction and augments incrementally by adding decision trees. Each subsequent tree is trained to model the residual errors produced by the current ensemble. Further, gradient descent is applied to minimize a specified loss function that quantifies prediction error. The algorithm also optimizes an objective function that integrates the empirical loss with explicit regularization terms.

9.3.2.3 Model Performance Evaluation

Given the presence of class imbalance in the dataset, multiple evaluation metrics are employed to obtain a comprehensive assessment of model performance, including accuracy, precision, recall, F_1-score, and the area under the ROC-AUC. The findings reported in [4, 6] indicate that decision tree-based models and XGBoost achieve superior performance in terms of both accuracy and F_1-score. Although gradient boosting produces comparable predictive results, XGBoost offers improved computational efficiency due to its optimized implementation.

RF models demonstrate robustness to noise and outliers, effectively capturing nonlinear relationships and higher-order feature interactions, and operating well across variables with differing scales without the need for feature normalization. While RFs yield moderate accuracy, their F_1-scores are comparatively lower, reflecting reduced sensitivity to minority class instances. Support vector machines exhibit stable and competitive performance, whereas the k-nearest neighbors' algorithm shows weaker predictive capability, particularly in identifying minority class observations.

To enhance transparency and better understand the factors influencing model predictions, SHAP-based interpretability analysis can be applied to examine feature contributions at both the global and instance levels.

9.3.3 Forecasting Financial Distress Using Deep Learning Models

The multifaceted nature of financial distress requires modeling frameworks capable of capturing nonlinear, high-dimensional, and dynamic relationships among diverse data sources. DL models address these requirements through layered architectures that acquire tiered illustrations from raw or minimally processed inputs [10, 11].

One of the key strengths of DL lies in its capacity to capture nonlinear relationships between financial indicators such as leverage, liquidity, profitability, and market volatility. These interactions often jointly influence financial distress but remain difficult to model using linear techniques. Deep neural networks apply nonlinear transformations at multiple levels, enabling the discovery of latent patterns embedded in complex financial data [10].

DL models are also well suited for high-dimensional datasets that combine firm-level financial ratios, market indicators, macroeconomic variables, and alternative data sources. By learning compact internal representations, DL models mitigate multicollinearity and redundancy, allowing scalable analysis as data dimensionality increases [3].

Furthermore, DL facilitates the incorporation of diversified data modalities. Structured numerical data can be treated alongside unstructured textual data, such as annual reports and news articles, within unified architectures. This multimodal capability enables a comprehensive assessment of corporate financial health by incorporating both quantitative performance metrics and qualitative sentiment indicators [13].

Temporal dependency is another defining characteristic of financial distress. Corporate deterioration typically unfolds over time, with early warning signals embedded in historical trends. DL architectures designed for sequential data effectively capture such temporal dynamics, enhancing early detection capabilities [11]. An overview of various DL models is given below:

A. **Feedforward Neural Networks:** Multilayer perceptrons (MLPs) are widely used as baseline DL models for financial distress prediction. They process structured numerical data through dense layers with nonlinear activation functions, enabling the modeling of complex feature interactions. Although MLPs do not explicitly model temporal dependencies, they serve as effective feature learners and are often integrated into hybrid frameworks [10].
B. **Recurrent Neural Networks and** Long Short-Term Memory **(LSTM)/**Gated Recurrent Unit: Recurrent neural networks are designed for sequential data analysis and uphold past observations. LSTM and gated recurrent unit architectures address vanishing gradient issues and enable long-term dependency learning. These models are particularly effective for analyzing multiyear financial statements and market time series, making them well suited for distress prediction [11].
C. **CNNs:** CNNs extract local patterns using convolutional filters. In financial applications, CNNs can be applied to transformed financial matrices or ratio trend repre-

sentations to capture localized patterns indicative of deteriorating financial conditions. Their parameter-sharing mechanism improves computational efficiency and generalization [14].

D. **Attention Mechanisms and Transformers:** Attention mechanisms allow models to focus selectively on relevant features or time periods. Transformer architectures, based entirely on self-attention, enable efficient modeling of long-range dependencies and parallel processing. In financial distress prediction, transformers have shown promise in analyzing lengthy financial disclosures and long-term time series [13].

9.3.3.1 Model Architecture

The model architecture is designed to process heterogeneous data sources within a unified predictive framework. A baseline DL architecture typically consists of parallel input branches that handle different feature types. Structured numerical features are passed through fully connected dense layers to learn nonlinear interactions among financial ratios. Time-series financial data are processed using recurrent architectures such as LSTM or bidirectional LSTM networks to capture temporal dependencies across reporting periods. Textual embeddings derived from transformer-based language models are further refined through attention or transformer layers to emphasize contextually important information. The outputs of these subnetworks are merged and passed through additional dense layers, culminating in a sigmoid-activated output neuron that estimates the probability of financial distress. Binary cross-entropy loss is commonly employed to optimize classification performance.

9.3.3.2 Forecasting Financial Distress Using Hybrid Models

Hybrid modeling integrates multiple analytical paradigms to leverage their complementary strengths. In financial distress prediction, hybrid frameworks commonly combine DL for representation learning with classical statistical or ML models for interpretability and decision-making [12].

One major advantage of hybrid models is improved predictive performance. By combining insights from multiple models and data sources, hybrid systems reduce information loss and enhance robustness across different economic conditions. This integration is particularly valuable during periods of economic instability, when single-model assumptions may fail [12].

Interpretability is another critical benefit. Hybrid frameworks often introduce transparent decision layers – such as logistic regression or survival analysis – on top of deep-learned features. This design supports regulatory compliance and stakeholder trust while retaining the expressive power of DL [1].

9.3.4 Hybrid Modeling Framework

The hybrid modeling framework integrates DL with classical statistical and ML techniques to balance predictive power and interpretability. In the first stage, deep neural networks are trained to learn high-level representations from structured, temporal, and textual data. These learned features are then fused into a single feature space, often followed by dimensionality minimizing techniques such as autoencoders or PCA to improve efficiency. In the next stage, traditional classifiers or survival analysis models – including logistic regression, RFs, Cox proportional hazards models, or gradient boosting – are applied to the fused representations. An ensemble layer may further combine predictions from multiple models using stacking or weighted voting schemes. This hybrid design leverages DL's ability to model complex nonlinear patterns while retaining the transparency and robustness of classical approaches.

9.3.4.1 Hybrid Models

A. **DL-Econometric Hybrid Models:** In DL-econometric hybrids, deep neural networks extract concealed features from economic data, which are subsequently used as inputs to interpretable econometric models such as logistic regression or Cox proportional hazards models. This approach enables nonlinear feature learning while conserving probabilistic interpretation and time-to-distress estimation [12].
B. **Ensemble-Based Hybrid Approaches:** Ensemble hybrids combine predictions from multiple ML and DL models using stacking, boosting, or weighted averaging. Such approaches reduce variance and bias, improving stability and performance on imbalanced financial datasets [12].
C. **Text and Structured Data Fusion:** Hybrid models increasingly integrate textual sentiment features derived from NLP models with structured financial ratios. This fusion enables the capture of both quantitative signals and qualitative context, resulting in more comprehensive distress prediction systems [13].

9.3.4.2 Hybrid Models Designed for Financial Assessment

This section presents some of the research and development done for financial distress estimation using hybrid techniques.

A hybrid ML framework for financial distress prediction [7], can be developed by identifying key variables with stepwise regression (SR) and the least absolute shrinkage and selection operator (LASSO). Subsequently, financial distress forecasting models can be developed by implementing classification and regression trees (CARTs) and RFs. SR is a statistical variable selection technique used in regression modeling to obtain a subset of explanatory variables that contribute to predicting a response variable and is expressed by eq. (9.6). It iteratively adds and removes predictor variables from a regression model based on predefined statistical criteria. This reduces dimensionality and enhances interpretability:

$$\hat{\beta} = \arg\min|\beta| \sum_{i=1}^{n} (y_i - X_i\beta)^2 \tag{9.6}$$

LASSO is a penalized regression technique that achieves concurrent variable selection and coefficient shrinkage by applying an L_1-norm constraint. This is expressed by eq. (9.7), where controls the degree of shrinkage:

$$\hat{\beta} = \arg\min|\beta| \left(\sum_{i=1}^{n} (y_i - X_i\beta)^2 + \lambda \sum_{j=1}^{p} (\beta_j) \right) \tag{9.7}$$

Two decision-tree algorithms that are used are CART and RF. After data collection and preprocessing, screening of key variables is performed by LASSO and SR for classification models, followed by four hybrid prediction models, namely LASSO-CART, LASSO-RF, SR-CART, and SR-RF. The LASSO-CART method yields higher accuracy followed by SR-CART. Similarly, LASSO-RF yields higher accuracy as compared to SR-RF model [7].

An MLP-ANN has also been suggested [8]. The financial status of firms can be classified based on whether bankruptcy probability is high or low. The input layer of MLP-ANN is composed of five Altman's *Z*-score-based explanatory variables. The neural network was built using 70% of the data sample and remaining 30% was utilized for prediction.

Harmony search algorithm with the optimal long short-term memory (HSA-OLSTM) model is another hybrid algorithm designed for financial crisis prediction of a company [9]. It employs HSA to identify optimal feature subsets and enhance the classification performance of the twin support vector machine. It explores the search space by generating harmonized solution vectors rather than relying on random search strategies. It performs LSTM-based classification for proper class labels allocation. This is followed by tuning variables using improved bat algorithm which overcomes early convergence issues. This hybrid optimized model shows higher accurate results, precision and recall as compared to tradition models like SVM.

9.4 Challenges and Considerations

Despite their strengths, hybrid DL models face several challenges. Financial distress datasets are often highly imbalanced, requiring techniques such as oversampling, class weighting, or focal loss to address bias. Interpretability remains a concern, particularly for deep components, necessitating explainability tools such as SHAP or LIME. Overfitting can occur due to model complexity and limited distress samples, highlighting the need for careful validation and regularization. Furthermore, the computational demands of transformer and recurrent architectures may require specialized hardware, increasing deployment costs.

9.5 Conclusion

DL combined with hybrid modeling presents a powerful framework for augmenting the accuracy of forecasting financial distress studies. By capturing complex nonlinear, temporal, and contextual patterns and integrating diverse data modalities, hybrid models overcome the limitations of traditional approaches. The synergy between deep representation learning and classical statistical inference enables robust, interpretable, and scalable distress prediction systems. Such frameworks hold significant potential for supporting early warning mechanisms and informed decision-making among investors, financial institutions, and regulators.

References

[1] Ha, H., Hung, D., & Tran, M. (2023). Financial distress forecasting with a machine learning approach. *Corporate Governance and Organizational Behavior Review*, *7*(2023). doi: 10.22495/cgobrv7i3p8

[2] CHAN, T., Choong, Y., & Nee, C. (2011). Factors affecting financial distress: The case of Malaysian public listed firms. *Corporate Ownership and Control*, *8*, 345–351. doi: 10.22495/cocv8i4c3art3

[3] The analysis of Altman model and Ohlson model in predicting financial distress of manufacturing companies in the Indonesia Stock Exchange (2017). *Indian-Pacific Journal of Accounting and Finance*, *1*, 51–63.

[4] Ruolin, Q. (2025). Enterprise financial distress prediction based on machine learning and SHAP interpretability analysis. In *Proceedings of the 2025 International Conference on Artificial Intelligence and Digital Finance (AIDF '25)* (pp. 76–79). New York, NY, USA: Association for Computing Machinery. https://doi.org/10.1145/3764727.3764740

[5] Duarte, D. L., & Barboza, F. L. D. M. (2020). Forecasting financial distress with machine learning – A review. *Future Studies Research Journal: Trends and Strategies*, *12*(3), 528–574. https://doi.org/10.24023/FutureJournal/2175-5825/2020.v12i3.533

[6] Kristanti, F. T., Salim, D. F., & Ihsan, A. F. (2025). Financial distress prediction in emerging markets by using a machine learning approach for financial sustainability. *Discover Sustainability*, *6*(1296). https://doi.org/10.1007/s43621-025-02199-1

[7] Chen, S., & Shen, Z.-D. (2020). Financial distress prediction using hybrid machine learning techniques. *Asian Journal of Economics, Business and Accounting*, 1–12. doi: 10.9734/ajeba/2020/v16i230231

[8] Nguyen, T. K. O., Truong, C. D., Nguyen, V. D., Buertey, S., Pham, H. T., Nguyen, H. L. A., & Hoang, T. N. (2024). Hybrid model for financial distress prediction: Evidence from vietnamese listed firms. In *Proceedings of the 2024 7th International Conference on Information Management and Management Science (IMMS '24)* (pp. 316–322). New York, NY, USA: Association for Computing Machinery. https://doi.org/10.1145/3695652.3695686

[9] Elhoseny, M., Alhashmi, S., Deif, M., Boujlil, R., & Metawa, N. (2025). Hybrid deep learning for bankruptcy prediction: An optimized LSTM model with harmony search algorithm. *Journal of Intelligent Systems*, *34*, 20250058. doi: 10.1515/jisys-2025-0058

[10] Elhoseny, M., Metawa, N., Sztano, G., & E-hasnony, I. M. (Feb. 2025). Deep learning-based model for financial distress prediction. *Annals of Operations Research*, *345*(2), 885–907. Springer. doi: 10.1007/s10479-022-04766-5

[11] Chen, W., Hussain, W., Cauteruccio, F., & Zhang, X. (2024). Deep learning for financial time series prediction: A state-of-the-art review of standalone and hybrid models. *Computational Modeling in Engineering and Sciences*, *139*(1), 187–224. techscience.com. doi: 10.32604/cmes.2023.031388

[12] Gao, R., Cui, S., Wang, Y., et al. (Jan. 2025). Predicting financial distress in high-dimensional imbalanced datasets: A multi-heterogeneous self-paced ensemble learning framework. *Finance-Innovations*, *11*, 50.

[13] Predicting financial distress using multimodal data: An attentive and regularized deep learning method. (Jul. 2024). *Information Processing and Management*, *61*(4). ScienceDirect. doi: 10.1016/j.ipm.2024.103703

[14] Shetty, S., Musa, M., & Brédart, X. (2022). Bankruptcy prediction using machine learning techniques. *Journal of Risk and Financial Management*, *15*(1), 35. MDPI. doi: 10.3390/jrfm15010035

[15] Liang, T., Lu, C. C., Tsai, C. F., & Shih, G. A. (Apr. 2024). Predicting financial distress using machine learning approaches: Evidence from China. *Journal of Contemporary Accounting and Economics*, *20*(1), 100403. ScienceDirect. doi: 10.1016/j.jcae.2024.100403

[16] Chinonyerem, S., et al. (Jun. 2025). Corporate bankruptcy and financial distress prediction using data analytic and machine learning techniques. *Asian Journal of Advanced Research and Reports*, *19*(6), 65–78. Asian Journal of Advanced Research.

[17] Survey, classification and critical analysis of the literature on corporate bankruptcy and financial distress prediction. (Mar. 2024.). *Machine Learning Applications*, *15*(100527). ScienceDirect. doi: 10.1016/j.mlwa.2024.100527

Ajay Sudhir Bale, Mamta B. Savadatti*, Varun Yogi Shanthakumar, Nikhil S. Nandi, Omkar K. Kanetkar

Chapter 10 AI-Driven Optimization Techniques for Intelligent Control Frameworks

Abstract: This chapter studies the transformative impact of artificial intelligence (AI)-driven optimization techniques on different intelligence control frameworks across domains such as industrial automation, energy systems, transportation, healthcare, and environmental management. The study examines how decision intelligence systems are utilized in AI-driven environments and how they enhance the resource management, their system efficiency. It should be operational stability in different environments. This work uses various case studies including Tesla's auto pilot system, Google DeepMind's data center cooling optimization, Siemens' smart factory production scheduling, GE's wind farm output maximization, and Honeywell's Heating, Ventilation, and Air Conditioning (HVAC) systems for commercial buildings to demonstrate how AI capabilities can be dynamically leveraged used to enhance the systems and their interactions with static models. The study also highlights how AI-powered intelligent structures can solve traditional limitations in scalability and adaptability and real-time adaptability. Specifically, the integration of decision-making processes, including multi-objective optimization, reinforcement learning, metaheuristic methods, multi-agent systems, swarm intelligence, and human-in-the-loop approaches, is shown to deliver enhanced agility, efficiency, and resilience across industrial applications while addressing broader sustainability goals.

Keywords: Control frameworks, efficiency, intelligent systems, industrial applications

10.1 Introduction

Decision intelligence (DI) is essential for the management of many systems. It entails using data, computations, and modeling to arrive at informed and optimal decisions

***Corresponding author: Mamta B. Savadatti**, Department of ECE, New Horizon College of Engineering, Bengaluru 560103, Karnataka, India, e-mail: mamta.savadatti@gmail.com
Ajay Sudhir Bale, Department of ECE, New Horizon College of Engineering, Bengaluru 560103, Karnataka, India
Varun Yogi Shanthakumar, Department of Electrical and Electronic Engineering, California State University, Sacramento, CA, USA
Nikhil S. Nandi, Department of Mechanical Engineering, CMR University, Bengaluru, Karnataka,India
Omkar K. Kanetkar, Department of ECE, New Horizon College of Engineering, Bengaluru 560103, Karnataka, India

 | https://doi.org/10.1515/9783112211991-010

in complicated and unpredictable situations. In smart logistics, DI enhances routing and distribution of resources, hence decreasing operating expenses and delivery times. In energy networks, it facilitates predicting demand and the incorporation of renewable energy (RE), hence maintaining grid stability and reliability. Aerospace systems utilize DI, especially in predictive maintenance, which improves safety and reduces maintenance. The integration of DI converts raw data into valuable information, facilitating real-time, precise, and tactical choices. This capacity enables organizations to manage intricacy, adjust to shifts, and sustain superior operation [1, 2]. Conventional decision-making systems encounter considerable difficulties when utilized in intricate commercial systems. Such approaches frequently rely on static architectures, rule-driven strategies, and rigid operational mechanisms, which do not adequately handle the dynamic and interconnected characteristics of contemporary systems [3, 4]. A significant drawback is the incapacity to consider real-time connections among systems elements. For example, in energy grids, static techniques can fail to accommodate abrupt variations in power generation or demand, for as breakdowns or inconsistent RE supply. Likewise, in logistics, static models are inadequate for managing disturbances such as traffic jams, port interruptions, or changing market trends [5, 6]. Scalability presents an additional complexity. As systems expand in scale and intricacy, conventional architectures become increasingly inefficient and vulnerable to errors. For instance, rules-driven management in aerospace system may neglect nuanced correlations in sensor data, resulting in unanticipated malfunctions [7, 8]. Moreover, these architectures frequently exhibit a deficiency in predictive capabilities, constraining pre-emptive reactions to disturbances. To address these drawbacks, decision-making procedures need to evolve to be particularly dynamic, data-driven, and adaptable to the intrinsic diversity and changing conditions of contemporary industrial operations [9]. Incorporating artificial intelligence (AI) enables industrial systems to attain enhanced responsiveness, effectiveness, and durability. AI-driven DI enables proactive answers to difficulties, surpassing conventional methods and providing exceptional organizational advantages.

The following sections discussed various technologies using AI as a solution to solve many problems in cross-domain applications.

10.2 Applications of Intelligent Control Systems Using AI

The different applications such as industrial automation, energy systems, transportation and mobility, healthcare, environmental management, smart infrastructure, financial systems, telecommunications, agriculture (precision farming), consumer electronics, and Internet of Things (IoT) are discussed in this section.

10.2.1 Industrial Automation

The rapid advancement of the production sector, propelled by the incorporation of advanced tools like AI, the IoT, and automated processes, has resulted in the advent of Industry 5.0, marking a significant transition from Industry 4.0 [10]. Industry 4.0 prioritizes the digitalization of production through information technology (IT) and smart devices that are interconnected to enhance production and efficiency, whereas Industry 5.0 highlights cooperation between humans and machinery, emphasizing long-term sustainability, customization, and social good. This new age aims to reconcile the automation and intellect offered by AI with the innovative thinking and ethical considerations associated with human labor, promoting a people-centered approach to production. Emerging Industry 4.0 technologies such as blockchain, artificial intelligence, and interconnected cyber-physical systems are increasingly being integrated to improve transparency, traceability, and intelligent decision-making in advanced manufacturing environments [11]. With worldwide consumers increasingly seeking personalized goods and ethical procedures, the significance of optimization in Industry 5.0 has grown essential [11, 12]. Optimization within Industry 5.0 entails addressing broad challenges that include diverse factors, including productivity, managing energy, ecological responsibility, and item customization. Due to the intricacy of contemporary industrial processes, conventional optimization techniques are inadequate. Sophisticated AI-driven optimization approaches are essential for managing the dynamic characteristics of such systems, which need consideration of real-time data, variable requests, and complex manufacturing steps [13, 14]. These strategies not only support decision-making but also empower producers to concurrently attain numerous, sometimes contradictory objectives, such as cost reduction, superior effectiveness, and minimal ecological impact [10]. Industry 5.0 underscores the integration of human intelligence and machine proficiency, prompting the refinement and augmentation of conventional optimization methodologies to satisfy emerging demands. The following essential methodologies are important in the advancement of optimization within Industry 5.0 (Figure 10.1):

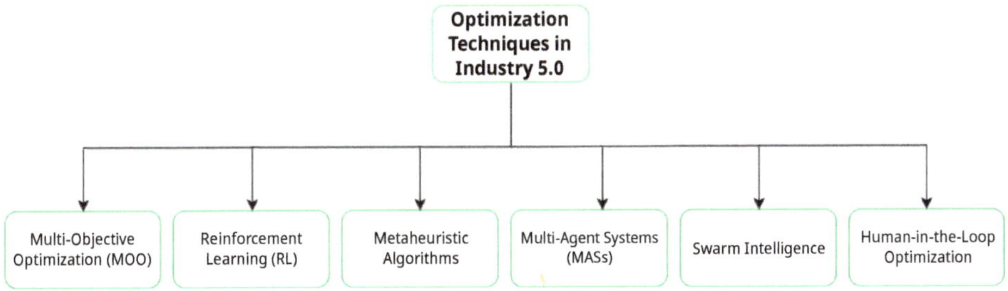

Figure 10.1: Principal optimization strategies in Industry 5.0 [7]. Reproduced from open access article.

- Multi-Objective Optimization (MOO): In the context of Industry 5.0, characterized by varied and frequently competing goals (e.g., maximizing productivity, minimizing energy consumption, and improving employee well-being), MOO methodologies such as Pareto optimization, non-dominated sorting genetic algorithm-II, and multi-objective particle swarm optimization are indispensable. These strategies enable individuals to assess many criteria and identify alternatives that provide optimal trade-offs among conflicting goals.
- Reinforcement Learning (RL): AI-based RL will be deployed in Industry 5.0 settings for continuous optimization. RL algorithms acquire knowledge via environmental interaction, progressively enhancing their decision-making capabilities depending on response. This method is especially effective for enhancing operations that need adaptive responses in unpredictable, swiftly evolving environments, such as modifying plans for manufacturing according to variable consumer preferences or optimizing resource distribution in real time.
- Metaheuristic methods, such as genetic algorithms (GAs), simulated annealing (SA), and ant colony optimization (ACO), are being enhanced to address the intricate, high-dimensional optimization challenges faced in smart factories (SFs). These heuristic methods are intended to efficiently navigate extensive solution spaces and may be tailored to tackle the broad complexities of Industry 5.0 issues.
- Multi-agent Systems (MASs): Distributed optimization is becoming more pertinent in Industry 5.0, where many subsystems inside an intelligent plant may possess competing objectives. MASs include individual agents, each tasked with optimizing a particular facet of the production process. They interact and cooperate to accomplish global optimization, guaranteeing that the production facility functions effectively as a unit while accounting for localized restrictions.
- Swarm Intelligence (SI): Drawing inspiration from the behaviors of social organisms such as ants and bees, SI methodologies, including particle swarm optimization (PSO) and ACO, are used to address intricate optimization challenges in Industry 5.0 contexts. These strategies are especially efficacious in scenarios wherein multiple participants (e.g., robots or machines) must work together to attain a shared goal, such as minimizing wasted materials or decreasing manufacturing lead times.
- Human-in-the-Loop Optimization: An essential element of Industry 5.0 is the Human-in-the-loop methodology, whereby employees and AI systems engage in real-time collaboration. Optimization algorithms aim to improve human decision-making by offering suggestions derived from extensive datasets, while people retain oversight of strategic decisions. This method guarantees the preservation of the moral, interpersonal, and intellectual aspects of production while using the computational capabilities of AI.

10.2.2 Energy Systems Using AI

Currently, the advancement of human civilization and its critical need for make energy systems essential in all societal facets, particularly in the household sector, industry, and transport [15]. Among the critical concerns pertaining to energy systems are their capacity to adjust to market and supply conditions, achieve optimal efficiency, and minimize impacts on the environment. Given the population growth and the need to meet energy demands for enhanced well-being and comfort, alongside the growing reliance on oil and gas and their detrimental environmental impacts, these concerns warrant special attention [16]. The use of RE and systems to mitigate detrimental ecological effects, enhance economic opportunities, and ensure operational safety is a viable approach to address these problems. Accurate understanding of the performance characteristics of these devices and their significant output characteristics is essential for optimum and practical use [17]. Due to the significant impact of environmental factors on renewable systems, it is crucial to use methodologies and models to predict these variations and enhance system efficiency and energy management. The requirement for an instrument to comprehend the relationships among multiple system variables and to fully use this data is crucial. To ascertain the power production of a wind system, it is essential to forecast the direction and speed of the winds in a given region [18]. To assess the output capacity of solar panels and wind power plants, it is essential to forecast the strength and direction of solar radiation in a given region. Additional examples are discussed in their corresponding sections. Moreover, the steady rise in global data generation during the latter half of the twentieth century, coupled with the substantial information generated through smart devices due to advances in technology, including the integration of the IoT into electrical systems, renders data-driven models highly efficient. Concerning the escalation of data generation, it is notable that in 2013, the global production, utilization, and interchange of data amounted to 4.4 ZB; however, this figure increased almost tenfold by 2019. According to the research, models are classified into two distinct groups: knowledge-driven and data-driven, for the purpose of predicting and optimizing power systems. Knowledge-based approaches are often formulated from a comprehensive grasp of processes and their processes, necessitating extensive expertise and information, which increases the likelihood of inaccuracy [19]. Data-driven methodologies, particularly those utilizing AI as a subset, are advantageous as they do not require experience or a comprehensive mathematical model. They can be effectively employed with suitable data and a general understanding of the system, garnering considerable attention. Currently, the issue of insufficient data is almost resolved, resulting in increased research interest and focus in this area [10]. Consequently, the programs and procedures of the latest and most effective models based on data and AI, referred to as machine learning (ML) and deep learning (DL), have gained significant popularity and widespread utilization. Generally, while addressing AI-related problems, we consider two perspectives. When considering a temporal variable in relation to the data presented in the problem, it is essential to analyze the issue via the lens of time series (TS), for which the applicable algorithms are documented in the pertinent part. However, if the dimension of time is irrelevant, only primary and standard ML techniques can be utilized [20]. The temporal dimension is crucial for forecasting essential characteristics; hence, the model must include both traditional and fundamental ML methods, as well as TS algorithms, to get precise outcomes. In the context of renewable systems, the rising prevalence of RE necessitates the representation of time-varying input data in energy system optimization analyses. In the last decade, TS aggregation has arisen to mitigate this difficulty by reducing the number of models.

Photovoltaic (PV) energy is a crucial and extensively accessible RE source. In light of the energy crisis and the imperative to safeguard the environment, investments by governments and corporations in this sector are escalating annually, particularly concerning AI uses for PV systems. PV systems have become popular in-home usage throughout Poland and Europe [17], driven by increasing ecological issues and the rising costs of fossil fuels. Energy management (EM) solutions for residential systems throughout Poland are crucial for reducing energy expenses and optimizing revenues. The novel AI-driven functionalities of the PV grid encompass bidirectional digital communication (comprising reports, advising, and control signals), self-management and self-healing capabilities (when required), adapting and changing of operational modes, adaptive control, protective and safety features, power quality management, voltage and frequency regulation, and the maintenance of system reliability (Figure 10.2). Data-driven (ML) models, particularly artificial neural networks (ANNs), are essential at all levels (local, network) since they enable the separation of control logic from system structure system, eliminating the need for very precise physical models. The broader use of AI will streamline discussions and diminish the intricacy of the model.

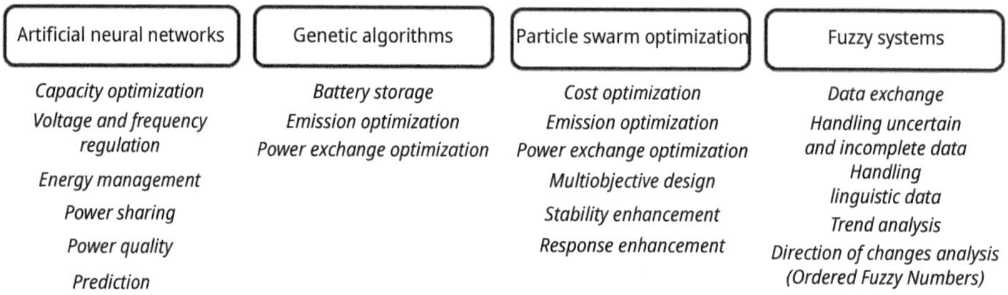

Figure 10.2: Contemporary and anticipated utilizations of essential AI innovations in PV grids [14]. Reproduced from open access article.

The study [17] analyzes the outcomes of several operating situations, focusing on two prosumer billing systems in Poland: net metering and net billing. The innovation and contributions lie in demonstrating that, for household PV networks, even rudimentary AI systems can effectively enhance inference and prediction to facilitate energy flow administration, making it less unpredictable and economical.

10.2.3 Transportation and Mobility

Transportation issues arise when the infrastructure and its users are too complex to model and forecast travel habits. Consequently, AI is considered well-suited to transport systems to address the challenges of rising passenger demand, CO_2 emissions, safety issues, and damage to the environment. These issues stem from the continuous increase in rural and urban congestion attributable to the growing population, particularly in emerging nations. In Australia, the expense of traffic is projected to reach 53.3 billion when the population rises to 30 million by 2031. In Melbourne, Australia, over 640 km of arterial highways experience congestion during peak hours, resulting in annual CO_2 emissions of 2.9 tons. Numerous twenty-first-century scientists attempt to establish a more dependable transportation system that minimizes impact on individuals and the environment, by employing inexpensive and advanced AI methodologies. It has potential applications for road construction, drivers, road users, and vehicles. AI applications in transportation have been evolving and being implemented in various manners. The work in [21] research study intends to examine three primary cases: the use of AI in business decision-making, strategic planning, and management. Addressing the challenge of incessantly increasing demand against constrained transportation capacity is crucial. This entails enhanced use of precise prediction and identification algorithms designed to improve the forecasting of traffic density, illnesses, and incidents. Applications of AI designed to enhance public transit are also examined. The deployment of partially autonomous driving systems such as Tesla Autopilot has further highlighted the importance of ethical decision-making, system reliability, and safety validation in AI-driven transportation frameworks [47]. Public transit is considered a sustainable method of movement. A potential AI application in transportation is linked and autonomous cars, which seeks to improve productivity by decreasing the incidence of accidents on roads.

In many instances, comprehending the interrelations among the attributes of the transportation system proves challenging; thus, AI methodologies might be proposed as an intelligent remedy for such intricate systems that elude management via conventional approaches. A multitude of studies has shown the benefits of AI in transportation. An illustration of this is converting road traffic sensors into intelligent agents that autonomously identify accidents and forecast foreseeable traffic situations. Numerous AI methodologies, including ANNs, are used in transportation. ANNs may be used for road organization, public transit, congestion identification, and forecasting traffic conditions. Numerous transportation issues result in optimization challenges that need custom methods to facilitate computational analysis. They are sophisticated computing techniques known as metaheuristic algorithms. The GA exemplifies such algorithms. It is based on the principles of evolutionary biology. It addresses intricate optimization challenges grounded in the notion of "survival of the fittest" and serves as an effective instrument in urban design networks. SA is another method derived from the simulation of the metal annealing process. The ACO is an AI system inspired by the habits of actual ants as they navigate from their nest to a food source. An artificial immune system patterned after the human immunological system. Bee colony optimization (BCO) addresses a hybrid complicated optimization issue. ACO and BCO are components of SI systems. SI is an AI system inspired by the collaborative behavior of ants and bees to achieve an efficient solution. The sophisticated computational analytics of these systems can encapsulate uncertainty, imprecision, and ambiguous ideas; hence, these approaches are used for route optimization challenges in transportation. A further optimization tool is the fuzzy logic model (FLM). It is used to address shortest route optimization. The performance of the FLM was contrasted to that of the logistic regression model in the development of a path selection model, with FLM demonstrating superior efficacy. Consequently, advanced methodologies such as FLM, GA, ANN, and ACO are well-suited for prediction, reasoning, and adaptation. As a result, they are used to address optimization challenges associated with changing traffic conditions and occurrences. A distinct software paradigm, termed agent-based software engineering (ABSE), has been presented based on AI ideas. ABSE facilitates a dynamic methodology for discovering the shortest route via the establishment of multi-criteria and multiple scenarios. Additionally, neural networks (NNs) were used to combine the framework with the previously stated methods for enhanced outcomes.

Due to the incremental transportation enhancement, the essence of mechanized mobility was preserved in steam engines. They transformed transportation methods, enhancing speed on both land and sea. Railways were essential to industrialization, facilitating the movement of raw materials and produced commodities, which in turn influenced urbanization and industry. This era had rapid urban expansion, with railroads linking cities to stimulate the development of the industrial sector [22]. Their arrival transformed maritime transit and illustrated worldwide commerce and colonization endeavors. The accessibility of personal automobiles enabled the expansion of suburbs, allowing residents to reside further from urban centers throughout the subsequent era of development. The infrastructure modifications to automobile transportation substantially transformed the social and economic roles of cities [23, 24]. The rise in urbanization necessitated the development of public transportation systems, exemplified by the New York and London subways. These technologies established the groundwork for contemporary urban transportation options. Initially developed for military applications, GPS technology subsequently transitioned to digitalization and interconnected mobility. It allowed real-time position monitoring as well as optimized the routing of personal and commercial vehicles [25]. The innovative business approach of online services, such as ride-hailing and car-sharing, has transformed city life by providing individuals with a variety of convenient transportation options.

In this era, self-driving technology began to advance in autonomous cars, with corporations like Google, Tesla, and Uber significantly supporting autonomous vehicle (AV) research [26]. AVs have the potential for reduced accidents and increased production, although they introduce complex legal and ethical challenges. Furthermore, global smart city architectures started implementation using sensors, IoT devices, and AI-driven insights to regulate traffic and assess air quality to reduce energy usage. The aforementioned information fundamentally underpins the notion of a "smart city" within urban planning [27, 28]. Transportation 5.0 advocates for the use of electric cars to mitigate emissions and promotes shared mobility to decrease automobile ownership. The primary elements included in this concept are ride-sharing, car-sharing, and bike-sharing. This stage of ecological preservation prioritizes decarbonization. Government initiatives are being implemented to maintain low pollution levels and develop sustainable infrastructure, including electric vehicle charging stations. This is consistent with international climate objectives aimed at minimizing the carbon footprint in transport. The transition from Transportation 1.0 to Transportation 5.0 signifies an increasing recognition of the social, technical, and environmental influences on mobility [29]. Each phase builds upon its predecessors, using technology, electronic devices, robotics, and enhancements in environmentalism to advance contemporary civilization. Transportation 5.0 heralds a promising trajectory toward a more profitable, productive, and human-centric future in transportation [30].

10.2.4 Healthcare

Global health systems (HSs) are at a pivotal juncture, with rapidly escalating healthcare costs that have significantly exceeded GDP growth rates, jeopardizing the viability of HS [31]. The situation became rather clear with the onset of the COVID-19 pandemic and the conflict in Ukraine. The confluence of constrained finances, an ageing population, increasing chronic illnesses, and the pressure on healthcare institutions, which have historically failed to meet heightened demands for services availability and access, is evident. Furthermore, the COVID-19 pandemic is causing HS failures in certain nations, such as India, Brazil, and Indonesia. HS relies on robust disease prevention protocols and evidence-based care strategies to address needs and standardize techniques in accordance with factories' healthcare programs. The notion of a highly reliable organization underscores the supervision of its amenities by both an accountable care organization or a health maintenance organization. The prevalence of long-term illnesses in the USA is progressively rising; specifically, 60% of individuals are afflicted with one chronic condition, while 40% suffer from more than two, resulting in yearly healthcare expenditures of USD3.3 trillion. Furthermore, this scenario rapidly evolved with the emergence of the infections, initially identified in Wuhan, China, in 2019, which was officially named COVID-19 by the World Health Organization on February 11, 2020. Since then, healthcare has been experiencing a digital transition that will alter lots of the essential components of medical treatment. This disease may result from the immense strain imposed by COVID-19 on world HSs, infrastructures supply chains, and personnel. The epidemic compelled medical professionals to embrace digital technology. Significant underlying transformations transpired in the medical field following the pandemic. Current-generation patients exhibit active participation during medical decision-making, attributed to the heightened acceptability of virtual medical systems and related digital technologies. Nonetheless, significant obstacles may arise, and the solutions to address them will facilitate the transition to the next healthcare era. The encounters and requirements of patients drive developments in the healthcare industry. Their primary focus is the establishment of digitally enhanced physician-patient relations and ensuring the availability of patient-centric services worldwide. The deployment of modern electronic equipment is essential for enhancing customer happiness, enabling tracking, monitoring health condition, and improving medication adherence. These elements could prove more advantageous throughout the post-hospitalization phase employing electronic medical systems. Simultaneously, healthcare consumers are apprehensive about disclosing their confidential information; therefore, healthcare organizations must maintain client confidence by demonstrating openness, compassion, and repeatability in their offerings. The growing deployment of generative AI technologies in intelligent systems has also raised important concerns regarding privacy preservation, ethical data usage, and responsible AI governance across digital infrastructures [38]. The emergence of biomedical science, encompassing genetics, digital medicine, AI, and its subset, ML, serves as the foundation for the evolution of healthcare, necessitating a new labor force and standards of practice. Genome and other methods, which includes as biometrics, tissue science, and the vaccine sector, have the potential to enhance and revolutionize testing, treatments, care delivery, regeneration therapies, and personalized medicine frameworks.

Digital health solutions offer physicians with a comprehensive perspective on patient health by enabling utilization of patient data. They also enable individuals to get additional health statistics from their providers [32]. These have genuine chances to enhance treatment outcomes and effectiveness; nonetheless, there are apprehensions over the potential psychological effects, especially with the prevalent use of Social Media (SM) and Instant Messaging Applications (IMAs) by patients, the public, and professional. Moreover, aggregated data from different places, including health information systems, mobile phones, telecommunications, mHealth and telehealth services, medical imaging devices (MIDs), and other AI-driven medical technologies, generate substantial datasets that enhance the application of ML and AI in healthcare systems [33, 34]. This is achieved through the learning processes derived from the data collected from these places, encompassing data from studies, user experiences, and the analysis of extensive datasets. Moreover, electronic health records (EHRs) include a range of medical data. Health information may be integrated utilizing innovative AI technology to provide accurate information into patient treatment. AI is considered a viable option for processing large-scale healthcare datasets. Additionally, big data analysis empowers healthcare practitioners to enhance their clinical services by optimizing EHRs using analytical algorithms [35]. These analytics use advancements in AI to filter large datasets on several criteria for enhanced data analysis. The work in [31] seeks to elucidate the role of AI in healthcare, emphasizing the following key aspects (Figure 10.3): (i) healthcare imaging and testing, (ii) virtual treatment for patients, (iii) medical investigations and development of drugs, (iv) involvement of patients and compliance, (v) recovery, and (vi) additional managerial applications [36].

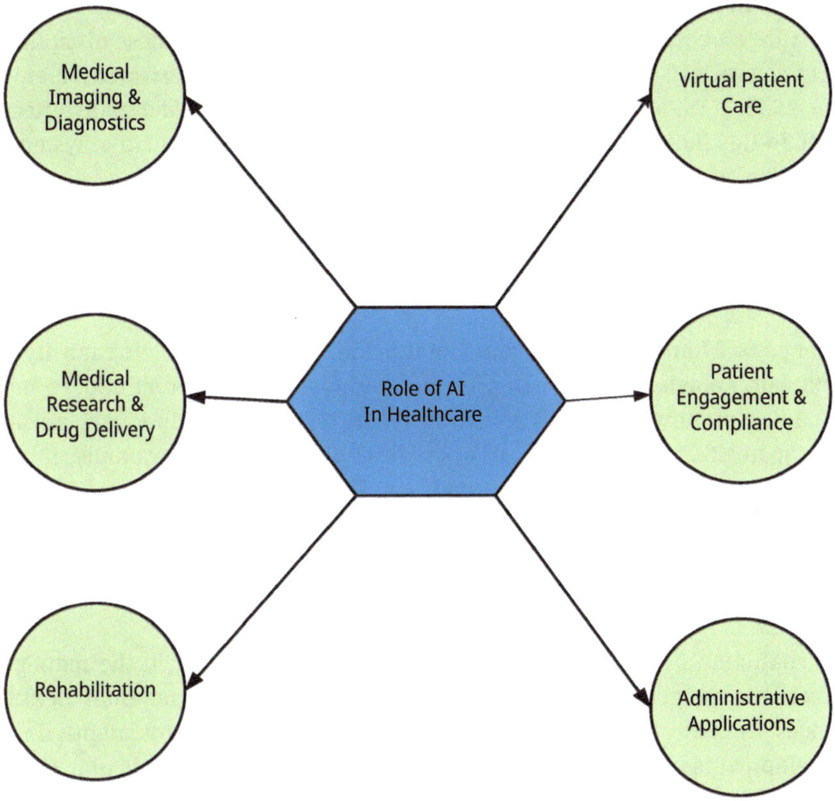

Figure 10.3: Utilization of AI in several facets of healthcare [28]. Reproduced from open access article.

10.2.5 Environmental Management

The advent of AI and DL technologies has signified a revolutionary era across several industries, especially in conservation [36]. These novel technologies have shown significant promise in advancing sustainable behaviors, particularly in RE and environmental health. The significance of AI and DL in reducing energy use and environmental deterioration is widely acknowledged. This section discusses the importance of AI in EM [37].

The RE business is undergoing a significant transition, mostly due to breakthroughs in AI and DL. These technologies have initiated a new epoch of economy and long-term viability generating considerable attention among scholars as well as business professionals. The progress and use of AI and DL in the RE industry may be distinctly classified into many key sectors [38]. Figure 10.4 demonstrates that these fields span a broad spectrum of functionality, including energy prediction and identifying anomalies in power systems, as well as more complex applications like RE system architecture and grid reliability. The interaction of AI and DL in these disciplines highlights their crucial significance in influencing the present and prospective environment of RE [39]. The graph presents a comprehensive picture of the many effects of AI and DL in this industry, clarifying their functions in energy predicting, systemic health evaluation, and other significant areas.

10.2.6 Environmental Health

The emergence of AI and DL in environmental health is driven by their capacity to analyze and gain knowledge from extensive datasets, yielding profound understanding and enabling prompt actions, as shown in Figure 10.5. The revolutionary advantages of AI for healthcare are shown across several disciplines. For example, in the realm of epilepsy, a common neurologic condition, DL approaches have markedly enhanced diagnosis processes [40, 41]. This automated approach, together with the combination of cloud computing and specific hardware for medical purposes, has enhanced the availability of computing resources and promoted an integrated scientific atmosphere [42].

The introduction of GeoAI signifies a significant advancement in the management of intricate geographical data within the realm of environmental health. GeoAI derives valuable insights from extensive geographical data by adeptly integrating developments in spatial science, AI, and DL [42–43]. GeoAI excels in processing large quantities of geographic and temporal data, guaranteeing efficient computing and providing the adaptability to modify methods and processes according to the distinct attributes of spatial activities. Furthermore, the flexibility of GeoAI allows the precise modeling of diverse environmental hazards across several geographic areas. Its promise includes quantifying previously mysterious circumstances, therefore enhancing the comprehension of the correlations between exposure to contaminants and illness. Thus, GeoAI is poised to transform environmental health, altering our methods of prediction, detection, and management of environmental health issues [44, 45]. The COVID-19 pandemic significantly underscored the crucial significance of AI, ML, and DL technologies in addressing global health emer-

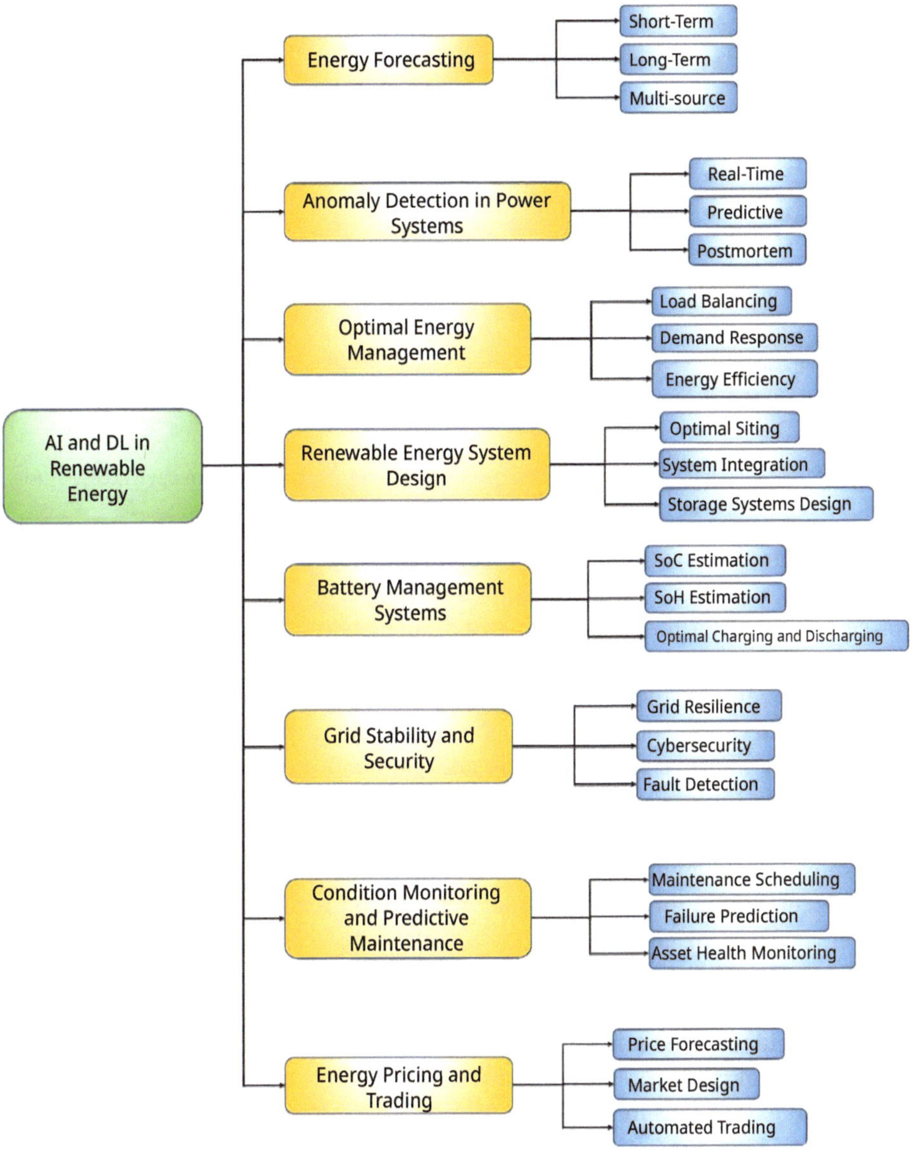

Figure 10.4: The use of AI and DL in RE; orange denotes several application settings; blue indicates sub-scenarios inside the above examples [33]. Reproduced from open access article.

gencies. There are many studies that provide critical insights into the use of AI-driven ML and DL techniques in the diagnosis and treatment of COVID-19 [46]. These procedures comprise a wide range, including noninvasive detection techniques, severity scoring evaluations, illness transmission modeling, and elements of

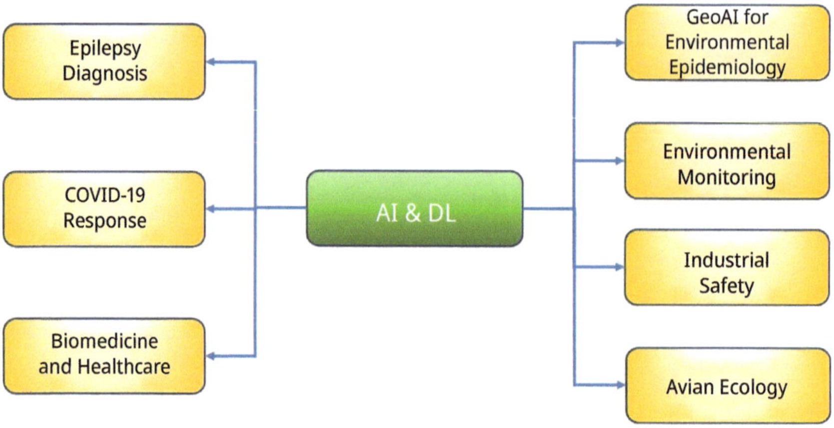

Figure 10.5: Utilization of AI and DL in environmental health [33]. Reproduced from open access article.

medicine manufacture and vaccine research. Moreover, these AI tools proved adeptly used in extensive study of data, including COVID-19 case data as well as online communication data, yielding essential insights into epidemic trends [47], routes of spread, and effects. Moreover, ML and DL have played a crucial role in epidemic prevention and public surveillance initiatives, such as airline security screenings, surveillance of patients, and epidemic identification. These developments not only improve our capacity to address the present epidemic but also facilitate the use of AI in preventing and taking control of upcoming healthcare crises. In the field of environmental monitoring, AI and DL have been used to enhance the forecasting of detrimental algal blooms [48]. Various studies have shown the efficacy of DL models in forecasting dangerous algal outbreaks in South Korea's principal rivers, surpassing conventional statistical analysis. The incorporation of AI and DL in ecological wellness has seen a significant increase in research activities, each offering distinct approaches and insights [49]. A comparison examination of current research elucidates the varied uses and the existing deficiencies in this field. A comparative study of current research is offered to enhance understanding of the different uses and extant gaps in this topic.

10.2.7 Smart Building Energy Management

The use of AI and DL in intelligent building EM has progressively gained momentum, leading to significant improvements in the forecasting, optimization, and automating of various EM activities, as shown in Figure 10.6 [36].

Utilizing the YOLOv3 algorithm, the technology can accurately ascertain the number of occupants in a designated area, hence facilitating optimum air conditioner operation in smart buildings. The nascent field of deep RL, a subset of AI, has significant promise in tackling EM issues in smart buildings. Digital twin methodologies may be essential in developing intelligent optimization and automation solutions for EM in residential districts. Their study highlights the effectiveness of solutions that integrate three-dimensional data models with IoT, AI, and ML to improve RE generation. EM efficiency in IoT networks may be attained with powerful DL architectures. This strategy utilizes several preprocessing methods to address the variability of power data, coupled with an effective decision-making algorithm for short-term planning on resource-limited devices. The interrelation between IoT-based real-time manufacturing logistics and cyber-physical process tracking devices is essential for overseeing the complexities and adaptability of SFs.

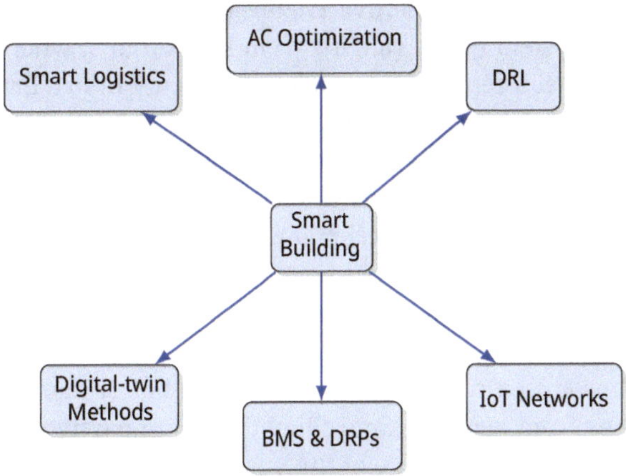

Figure 10.6: The transformation of AI and DL in the EM of smart buildings [33]. Reproduced from open access article.

10.3 Methodology

This section discusses various cases by bringing out the recent trend in the cross-discipline of AI and provides solutions to the many problems.

10.3.1 Tesla's Autopilot System Energy Management

Automobiles have significantly evolved since their inception over a century ago. Currently, in the USA, there exists a ratio exceeding one automobile for each licensed driver. This underscores the need for people to possess driving capabilities to engage successfully in the contemporary economy and have a fulfilling social life [47, 48]. Although driving facilitates economic engagement and social enhancement, it remains a primary cause of mortality in the USA. Annually, about 30,000 lives are lost on American highways, mostly due to human mistake. Recent technological advancements have resulted in the creation and implementation of advanced driver assistance systems (ADAS) aimed at improving safety and efficiency in driving. ADAS technologies not only improve driving safety but also function as foundational elements for the future creation of driverless cars. Society of Automotive Engineers (SAE) now delineates six tiers of Automated Driving System (ADS), ranging from levels 0 to 5, where level 0 signifies no automation and level 5 represents complete autonomy in driving. Partially ADS, designated as level 2 (L2) systems, are characterized by SAE J3016 for automated cars where the system executes steering and acceleration or deceleration, but the human driver supervises the conditions of driving. The driver must remain vigilant and focused to identify and react suitably to occurrences that exceed the limitations of the partly ADS. Consequently, for a partly automated car to function well, the driver must be vigilant at all times. Nonetheless, research indicates that drivers' capacity to perceive essential elements within the setting or to identify system alterations or malfunctions may be compromised compared to attentive and alert manual driving, as their function transitions from engaged controlling the car to passive observation, basically due to a decreased understanding of the system's status and procedures. Developers of partly autonomous driving (AD) cars now face the difficulty of ensuring that drivers comprehend their responsibilities and maintain vigilance while driving such vehicles. Specifically, there are apprehensions that drivers' propensity to partake in non-driving chores may be diminished when operating a partly AV, owing to the heightened utilization of portable electronic gadgets and the decreased burden afforded by these systems. Nevertheless, little research has been conducted on driver attitudes and behaviors while driving a largely AV. Due to much conjecture and little experience, accurately predicting consumer attitudes and behaviors regarding vehicle autonomy capabilities has proven challenging. Tesla is now one of the few manufacturers offering a car equipped with partly automated driving capabilities for the driver (i.e., autopilot). The objective of the current research was to comprehend the driving behavior of Tesla owners when autopilot is on and to elucidate the factors influencing the use of this technology by drivers. Naturalistic research was undertaken using instrumented Tesla automobiles to gather data over a 3-day duration for the assessment of these objectives.

Millions of vehicle accidents in the USA each year mostly stem from human mistake, resulting in 35,000 deaths and over $871 billion in damages. Tesla innovated in the realm of autonomy. Automotive solutions such as "autopilot" aim to reduce human errors and require for controllers in vehicles. The rapid progression of AD technology is transforming the public transit sector, providing enhanced convenience and safety while posing ethical challenges, as shown by the deadly encounter between Joshua Brown and Tesla's Autopilot system. Mr Brown's abrupt death on May 7, 2016, resulting from his Tesla Model S colliding with a tractor-trailer while in autopilot mode, highlights the ethical dilemmas linked to emerging technological advancements, particularly when human lives are at stake. Tesla documented the first fatality after over 130 million miles of autopilot use, a striking disparity compared to the roughly 94 million miles among deaths on US highways. The National Highway Traffic Safety Administration (NHTSA) commenced an evaluation, highlighting the need to ascertain if the equipment functioned as anticipated. Tesla swiftly informed the NHTSA of a distinct occurrence on a split highway in which a tractor-trailer across the road, causing the Model S to pass under the trailer and strike the windscreen. The fundamental concern in Joshua Brown's deadly accident with Tesla's Autopilot system is to the moral impacts of AD technology. Tesla Motors' ethical obligations in the implementation and promotion of a system necessitating continuous human oversight, in spite of its designation as Autopilot, evoke apprehensions over customer awareness, security features, and the ethical transition from assistance from drivers to partial automation. Nonethical factors, like technical and legal constraints, provide complexity, requiring a thorough examination of the autopilot system's functionality and regulatory structures [48].

10.3.2 Google DeepMind's Data Center Cooling Optimization

The rapid proliferation of Internet-enabled gadgets, together with the transition from consumer-side computation to Software as a Service (SaaS) and cloud-based services, is propelling the expansion of large-scale data centers (DCs). Propelled by substantial advancements in hardware cost and the exponential expansion of Big Data, the contemporary. An internet-based company comprises several attributes, such as customized experiences for clients and reduced downtime. Simultaneously, prominent hosting providers like Google Cloud Platform and Amazon Web Services have significantly reduced initial capital and operational expenses, enabling organizations with limited IT resources to expand rapidly and effectively to millions of subscribers. These changes are responsible for the emergence of massive distribution centers and associated problems with operations. Managing power is one of the most intricate difficulties. Rising energy expenses and ecological responsibility have intensified the demand on the DC sector to enhance its operating efficiency. As to Koomey, DCs accounted for 1.3% of world energy consumption in 2010 [49]. At this magnitude, even little enhancements in efficiency result in substantial cost reductions and save millions of tons of carbon emissions. Although it is well recognized that Google, alongside other prominent Internet corporations, has achieved considerable advancements in enhancing DC effectiveness, the general rate of power usage effectiveness (PUE) decrease has decelerated due to decreasing benefits and the constraints of current cooling technologies. Moreover, optimal methodologies, such as hot air containment, water-side economization, and thorough monitoring, have become standard in large-scale DCs. ML is ideal for the DC context due to the intricacy of manufacturing processes and the plethora of available data for monitoring. The contemporary large-scale DC encompasses a diverse array of pneumatic and electrical apparatus, together with their corresponding settings and management systems. The interplay among these components and several feedback loops complicates the precise prediction of DC performance by conventional engineering methods. A little adjustment to the cold-aisle temperature setpoints will result in load fluctuations within the cooling infrastructure (chillers, cooling towers, heat exchangers, pumps, etc.), subsequently leading to nonlinear alterations in machinery efficiencies. Ambient weather factors and instrument management will influence the resultant DC efficiencies. Employing conventional methods for predictive modeling often results in significant inaccuracies due to their inability to account for intricate interconnections. Moreover, the extensive array of potential device configurations and their corresponding setpoint values complicates the identification of maximum efficiency. In a live DC, achieving the desired setpoints may be accomplished by an assortment of devices (physical and electric apparatus) and software (control methodologies and setpoints). Evaluating all possible feature combinations to optimize efficiencies may prove impractical due to time limitations, variable IT loads and environmental factors, along with the necessity to continue a stable DC environment. An NN is used as the mathematical foundation for developing DC efficiency algorithms to tackle these difficulties. NNs are a category of ML techniques that emulate cognitive functions via interactions among artificial neurons. NNs are beneficial for modeling complex systems since they eliminate the need for users to establish feature conversations, hence inferring connections inside the data autonomously. The NN identifies patterns and interactions among features to autonomously create an optimal model. Typical applications of this ML branch are detection of voices, computer vision, and programming with autonomous robots. Similar to ML, the model's accuracy enhances with time with the acquisition of additional training data.

ML now enhances us in daily life via phone aids, picture recognition, and translation. However, technology may also assist us in addressing some of the worldwide most formidable physical challenges, like energy usage. Extensive industrial and manufacturing processes, such as DCs, exhibit significant energy consumption. Although considerable efforts have been made to mitigate energy use, much work remains due to the escalating global need for computing capacity. Over the last decade, we have prioritized the reduction of energy consumption: Google has developed own highly reliable servers, innovated more effective cooling methods for the DCs, and made substantial investments in RE sources, aiming for complete reliance on RE. Compared to 5 years ago, we now get almost 3.5 times the computational power from the equivalent energy expenditure, with ongoing annual changes. Significant advancements are rare; hence, we are pleased to announce that by using DeepMind's ML in Google DCs, they have achieved a reduction of up to 40% in energy consumption for cooling. In any extensive energy-consuming setting, this could be a significant enhancement. Considering the advanced nature of Google's data facilities, this is a remarkable advancement. The ramifications are substantial for Google's DCs, considering its capacity to significantly enhance energy usage and diminish emissions overall. This will also assist other organizations using Google's cloud in enhancing their energy efficiency. Although Google is but one of several DC operators globally, many do not use RE as they do. Enhancements in DC efficiency diminish overall emissions, and with technologies such as DeepMind's, ML is used to minimize energy consumption and tackle a significant challenge: climate change. Cooling is a key source of energy consumption in DC environments. Similar to the computer, DCs, which house the servers that operate Google Search, Gmail, YouTube, and other services, produce significant heat that must be dissipated to maintain optimal server functionality. This cooling is often achieved by substantial industrial apparatus, including pumps, chillers, and cooling towers. Nevertheless, dynamic settings such as DCs pose challenges to optimum operation for many reasons: The machinery involved in its operation, and surroundings interact in intricate, nonlinear ways. Conventional formulaic development and natural instincts often fail to encapsulate these connections. The framework is unable to rapidly adjust to either outside or inside fluctuations, such as meteorological conditions. This is due to our inability to formulate norms and heuristic in each operational circumstance. Every DC has a distinct architecture and atmosphere. A model specifically calibrated for a particular system might not be suitable for another. Consequently, a comprehensive intelligence architecture is essential for comprehending the interactions inside the DC. To resolve this issue, they began the use of ML 2 years ago to enhance the operational efficiency of DCs. In recent months, DeepMind developers collaborated with Google's DC team to enhance the system's value substantially. They developed an enhanced and adaptable approach to comprehend DC dynamics and enhance efficiency by using a set of NNs trained on various operational situations and characteristics inside the DCs. They achieved so through employing historical information previously gathered by several sensors inside the DC – data including temperatures, power levels, pump speeds, setpoints, and more – to train a group of DNN. The purpose was to enhance DC energy efficiency; hence, they trained the NNs on the projected average PUE [50], which is the ratio of overall facility energy consumption to IT energy consumption. Subsequently, they trained two further ensembles of DNNs to forecast the forthcoming temperature and pressure of the DC for the following hour. The objective of these assumptions is to replicate the recommendations derived from the PUE model, ensuring compliance with all operational limitations. They evaluated the methodology by implementing it in an actual DC.

10.3.3 Siemens' Smart Factory Production Scheduling

An SF is a completely digitalized manufacturing model that integrates product lifecycle management data with SCADA, PLCs, and other automation tools [51]. The key capabilities and innovations of SF for electronics firms are shop-floor connection, sophisticated robots, adaptable robotics, automation norms, virtual and augmented reality, and management of energy. SFs exhibit exceptional efficiency, conserving energy, time, and resources [52]. To achieve an unambiguous comprehension of the norms, designs, and use cases necessary for establishing smart manufacturing, RAMI4.0 was established. The design integrates essential components of I4.0 using a three-dimensional layer concept. Smaller businesses may adapt to Industry 4.0 by following the steps outlined in the standards; by applying even incomplete standards, they may be able to produce an early application for Industry 4.0. The identification of standards needed for applicable use cases is also made possible by RAMI 4.0 [53] and its architecture is shown in Figure 10.7

Working with companies to oversee and improve their business processes is something Siemens has done for a long time, all over the world. In order to effectively support employees, procedures, and technology, Siemens Smart Connectivity rings allow the convergence of the digital and physical worlds in production. By assisting its clients in selecting and deploying the most suitable administration systems and security measures to achieve efficiency, safety, resiliency, along with environmental objectives, Siemens is dedicated to assisting businesses on their path to the SF. The SF solution from Siemens encompasses the following services and technologies:

A cloud-based platform that lays the groundwork for services based on data is NAVIGATOR. In order to facilitate choices that lead to improved business results, it consolidates several statistics pertaining to energy usage, ecological variables, and asset health and lifespan. DESIGO is an application for building oversight that aims to improve workplace security, resilience, and economy. It provides a unified view of the manufacturing facility environment for managers by using accessible restful APIs to facilitate the flow of data from different systems and sensors. It does more than just connect gadgets and machines; it also optimizes procedures, satisfies unique facility needs, and makes it possible for devices to talk to each other and work together under one roof.

SIEMENS XCELERATOR is an all-inclusive computer service offering that includes a software building environment. It aids businesses in digitizing products that may be customized to meet the unique demands of customers and industries. BUILDING X is an open-platform set of AI-enabled apps that allow businesses to analyze water use patterns, emissions statistics, and utilization of energy in order to find efficient methods and optimize processes. In order to facilitate the integration of building information modeling, construction management system, IoT devices, and building management systems, ECODOMUS is a digital twin program that allows users to create digital models of resources and structures [54].

Figure 10.7: Rami 4.0 architecture [53]. Reproduced from open access article.

10.3.4 GE's Wind Farm Output Maximization

Increases in wind farm yearly energy production, improvements in service efficiency, and the introduction of new income streams for customers have all been achieved via the many software and hardware capabilities that General Electric (GE) has developed and commercialized when the Brilliant wind system was debuted in early 2013. Utilizing cutting-edge algorithms for control, data analysis, and power electronics/grid integration knowledge, GE is accomplishing more using less. This includes making better use of previous hardware, incorporating wind energy seamlessly around the grid despite its fundamental variability, navigating the intricacy of farms that contain numerous turbines, and generating wind profitable and predictable even in places with poor infrastructure. As we face the challenges and possibilities of the future, GE will continue to revolutionize the wind sector by using the industrial network to build breakthrough features inside the Brilliant wind platform [55]. GE has spent over $2 billion on research and development since 2002, when it first entered the wind industry. When it comes to wind power, GE has accomplished a lot in the last 12 years. In many parts of the globe, boosting the number of turbines from 85% to more than 98%, boosting capacity ratios from 25% to 50%, and reducing the cost of wind power to less than 5 cents/kwh. One of the most cost-effective options for generating new electricity is wind. Having addressed and resolved several crucial issues that have plagued the business from its start. GE has decided to pursue another opportunity: in light of the worldwide developments in data, communication, and computer systems. What other ways can we maximize the use of the gear we are installing? The "Industrial Internet" is made up of three parts: intelligent devices, which link individual units, fleets, and networked with managerial software and sophisticated sensors; complex analytics that make use of physics-based analysis, deep domain knowledge, and forecasting algorithms to comprehend and foretell the behavior of systems and devices; and people, who leverage these insights to make informed decisions and optimize operations. Launched in 2013, the Brilliant wind platform increases wind farm annual generation of electricity, increases services efficiency, and opens up new income streams for customers by using the energy of the Industry Web. Commercialization and implementation of several technologies from this platform have been carried out by GE since the debut.

10.3.5 Honeywell's HVAC System for Commercial Buildings

The use of various HVAC systems is becoming more important in maintaining interior temperatures in institutions such as colleges, hospitals, hotels, commercial workplaces, and industrial facilities, where the frequency and severity of severe weather events are becoming more common. Furthermore, a large portion of such activities are powered by antiquated systems that rely on fossil fuels such as natural gas. When structures are heated, conditions rise, and with them, the released particles of potentially dangerous carbon gases. There is a rising need for creative techniques that minimize energy expenses, provide pleasant working conditions, and fulfill the ever-increasing demands of federal, state, and national regulations concerning greenhouse gases and environmental goals. To better monitor and operate their boiler and burner environments, Honeywell Thermal Solutions offers a wide range of burner management technologies, gas valves, and flame detectors to HVAC clients. Through increased across the company system connection and accessibility, products and services improve pollution prevention and overall operational efficiency. Virtual control, analytical forecasting, and current system expertise are made possible by these advanced environmental devices, tools, and programs, such as HVAC tracking systems. This allows for improved decision-making and quicker action-taking. As older HVAC systems are expected to perform more and more, and as businesses investigate the possibility of implementing more modern technology, including heat pumps and electric boilers, this is crucial to oversee the heating landscape holistically [56].

10.4 Discussions and Future Scope

The implementation of AI-driven optimization techniques across various sectors has demonstrated significant potential for enhancing efficiency, sustainability, and operational performance. As evidenced by Google DeepMind's DC cooling optimization achieving 40% energy reduction and Tesla's advancements in AD systems, AI technologies are proving their capability to manage complex, interdependent systems with greater precision than conventional approaches. However, these advancements come with certain challenges that need addressing as the field evolves.

One critical area for future development is the ethical dimension of AI implementation in critical systems. The case of Tesla's autopilot system highlighted the ethical dilemmas associated with partially automated systems, where human oversight remains necessary but user attention may be compromised. As AI systems take on greater decision-making responsibilities in safety-critical applications, more robust frameworks for ethical AI deployment will be essential, balancing innovation with responsible implementation. Data quality and accessibility represent another significant challenge for the widespread adoption of AI in intelligent control frameworks. While large organizations like Google and Tesla have abundant data for training sophisticated models, smaller entities may struggle with data limitations. Future research should focus on developing transfer learning techniques and synthetic data generation methods that can function effectively with smaller datasets, democratizing access to AI-driven optimization benefits across organizations of all sizes. The integration of AI with human decision-makers remains an evolving area requiring further investigation. Industry 5.0 emphasizes human-machine collaboration rather than replacement, suggesting a need for AI systems designed to complement human intelligence rather than substitute it. Research into effective human-in-the-loop optimization approaches, as mentioned in the industrial automation section, will be crucial for maximizing the benefits of both human creativity and machine precision. Energy efficiency in AI systems themselves presents another important research direction. While AI can optimize energy usage in various applications, the computational requirements of sophisticated AI models can be substantial. Future work should explore more energy-efficient AI architectures and training methodologies to ensure that the energy savings achieved through optimization outweigh the energy costs of running the AI systems themselves. Interoperability between different AI systems and frameworks presents challenges in creating truly integrated intelligent control environments. The RAMI 4.0 architecture discussed in relation to Siemens' SF demonstrates one approach to standardization, but further efforts are needed to develop universal protocols and interfaces that allow seamless communication between diverse AI systems across organizational boundaries.

The resilience of AI-driven optimization systems against adversarial attacks and system failures requires additional attention. As these systems become more central to critical infrastructure, ensuring their security against manipulation and their graceful degradation in failure scenarios becomes increasingly important. Research into robust AI that can maintain functionality under suboptimal conditions will be essential for real-world implementations. Finally, regulatory frameworks for AI in control systems need further development to keep pace with technological advancements. Current regulations often lag behind innovation, creating uncertainty for organizations implementing cutting-edge AI solutions. Collaborative efforts between industry, academia, and regulatory bodies will be necessary to develop appropriate governance frameworks that encourage innovation while protecting public interests, particularly in domains like transportation and healthcare where human safety is directly impacted.

References

[1] Ekundayo, F. (2024). Leveraging AI-driven decision intelligence for complex systems engineering. *International Journal of Research Publications and Reviews*, *5*(11), 1–10.

[2] Biswas, P., et al. (2024). AI-driven approaches for optimizing power consumption: A comprehensive survey. *Discover Artificial Intelligence*, *4*(1), 116. Cdacd.

[3] Zohuri, B. (2024). Artificial intelligence and machine learning driven adaptive control applications. *Journal of Material Sciences and Engineering Technology*, *24*, 1–4. Cdacdac

[4] Oloruntoba, N. O. (2025). AI-Driven autonomous database management: Self-tuning, predictive query optimization, and intelligent indexing in enterprise it environments. *World Journal of Advanced Research and Reviews*, *25*(2), 1558–1580. .Cdcdscds.

[5] Xinfa, T., Jingjing, W., & Shengyuan, Z. (2025). *AI-Driven Integrated Optimization for Virtual Power Plants in Smart Grids.*

[6] Aldoseri, A., Al-Khalifa, K. N., & Hamouda, A. M. (2025). A framework for building resilience through innovation and process optimization in AI-powered digital transformation. In *Handbook of Digital Innovation, Transformation, and Sustainable Development in a Post-Pandemic Era* (pp. 3–33). CRC Press.

[7] Chen, S.-C., Chen, H.-M., Chen, H.-K., & Li, C.-L. (2024). Multi-objective optimization in industry 5.0: Human-centric ai integration for sustainable and intelligent manufacturing. *Processes*, *12*, 2723. https://doi.org/10.3390/pr12122723

[8] Vinay, N., Tiwari, S., & Baby, C. R. (2022). Artificial intelligence as a tool for conservation and efficient utilization of renewable resource. In Vyas, A. K., Balamurugan, S., Hiran, K. K., & Dhiman, H. S. (eds.). *Artificial Intelligence for Renewable Energy Systems.* https://doi.org/10.1002/9781119761686.ch2

[9] Zhao, Shuai, Frede Blaabjerg, and Huai Wang. "An overview of artificial intelligence applications for power electronics." IEEE Transactions on Power Electronics 36.4 (2020): 4633–4658

[10] Mohammadi, Ebrahim, et al. "A review on application of artificial intelligence techniques in microgrids." IEEE Journal of Emerging and Selected Topics in Industrial Electronics 3.4 (2022): 878–890.

[11] Purohit, T. P., Hashim, M. F., & Navale, S. (2022). Blockchain and its applications in industry 4.0. In Chatterjee, J. M., Garg, H., & Thakur, R. N. (eds.). *A Roadmap for Enabling Industry 4.0 By Artificial Intelligence*. https://doi.org/10.1002/9781119905141.ch16

[12] Forootan, M. M., Larki, I., Zahedi, R., & Ahmadi, A. (2022). Machine learning and deep learning in energy systems: A review. *Sustainability*, *14*, 4832. https://doi.org/10.3390/su14084832

[13] Mosavi, A., et al. (2019). State of the art of machine learning models in energy systems, a systematic review. *Energies*, *12*(7), 1301.

[14] Rojek, I., et al. (2023). Machine learning-and artificial intelligence-derived prediction for home smart energy systems with PV installation and battery energy storage. *Energies*, *16*(18), 6613.

[15] Szczepaniuk, H., & Szczepaniuk, E. K. (2022). Applications of artificial intelligence algorithms in the energy sector. *Energies*, *16*(1), 347.

[16] Lytras, M. D., & Chui, K. T. (2019). The recent development of artificial intelligence for smart and sustainable energy systems and applications. *Energies*, *12*(16), 3108.

[17] Franki, V., Majnarić, D., & Višković, A. (2023). A comprehensive review of artificial intelligence (AI) companies in the power sector. *Energies*, *16*(3), 1077.

[18] Abduljabbar, R., Dia, H., Liyanage, S., & Bagloee, S. A. (2019). Applications of artificial intelligence in transport: An overview. *Sustainability*, *11*, 189. https://doi.org/10.3390/su11010189

[19] Shamsuddoha, M., Kashem, M. A., & Nasir, T. (2025). A review of transportation 5.0: Advancing sustainable mobility through intelligent technology and renewable energy. *Future Transportation*, *5* (8). https://doi.org/10.3390/futuretransp5010008

[20] Lukic Vujadinovic, V., et al. (2024). Ai-driven approach for enhancing sustainability in urban public transportation. *Sustainability*, *16*(17), 7763.

[21] Shulajkovska, M., et al. (2024). Artificial intelligence-based decision support system for sustainable urban mobility. *Electronics*, *13*(18), 3655.

[22] Chen, G., & Wan Zhang, J. (2024). Intelligent transportation systems: Machine learning approaches for urban mobility in smart cities. *Sustainable Cities and Society*, *107*, 105369.

[23] Duggal, A. S., et al. (2021). Infrastructure, mobility and safety 4.0: Modernization in road transportation. *Technology in Society*, *67*, 101791.

[24] Mahapatra, S., Raj, S., & Badi, M. (2023). Smart transportation based on AI and ML technology. In *Artificial Intelligence and Machine Learning in Smart City Planning* (pp. 281–299). Elsevier.

[25] Iyer, L. S. (2021). AI enabled applications towards intelligent transportation. *Transportation Engineering*, *5*, 100083.

[26] Ushakov, D., et al. (2022). Artificial Intelligence as a factor of public transportations system development. *Transportation Research Procedia*, *63*, 2401–2408.

[27] Jiang, J., et al. (2024). Smart transportation systems using learning method for urban mobility and management in modern cities. *Sustainable Cities and Society*, *108*, 105428.

[28] Al Kuwaiti, A., Nazer, K., Al-Reedy, A., Al-Shehri, S., Al-Muhanna, A., Subbarayalu, A. V., Al Muhanna, D., & Al-Muhanna, F. A. (2023). A review of the role of artificial intelligence in healthcare. *Journal of Personalized Medicine*, *13*, 951. https://doi.org/10.3390/jpm13060951

[29] Bekbolatova, M., et al. (2024). Transformative potential of AI in healthcare: Definitions, applications, and navigating the ethical landscape and public perspectives. *Healthcare*, *12*(2), 1–31, MDPI.

[30] Sodhro, A. H., & Zahid, N. (2021). AI-enabled framework for fog computing driven e-healthcare applications. *Sensors*, *21*(23), 8039.

[31] Majnarić, L. T., et al. (2021). AI and big data in healthcare: Towards a more comprehensive research framework for multimorbidity. *Journal of Clinical Medicine*, *10*(4), 766.

[32] *Li, F.*, Ruijs, N., & Yuan, L. (2022). Ethics & AI: A systematic review on ethical concerns and related strategies for designing with AI in healthcare. *Ai*, *4*(1), 28–53.

[33] Fan, Z., Yan, Z., & Wen, S. (2023). Deep learning and artificial intelligence in sustainability: A review of SDGs, renewable energy, and environmental health. *Sustainability*, *15*, 13493. https://doi.org/10.3390/su151813493

[34] *Yu, L.*, et al. (2021). A review of deep reinforcement learning for smart building energy management. *IEEE Internet of Things Journal*, *8*(15), 12046–12063.

[35] Rocha, P., Siddiqui, A., & Stadler, M. (2015). Improving energy efficiency via smart building energy management systems: A comparison with policy measures. *Energy and Buildings*, *88*, 203–213.

[36] Selvaraj, Rajalakshmi, Venu Madhav Kuthadi, and S. Baskar. “Smart building energy management and monitoring system based on artificial intelligence in smart city.” Sustainable Energy Technologies and Assessments 56 (2023): 103090.

[37] Li, Yuanzheng, et al. “Artificial intelligence-based methods for renewable power system operation.” Nature Reviews Electrical Engineering 1.3 (2024): 163–179.

[38] Bale, A. S., et al. (2024). The impact of generative content on individuals privacy and ethical concerns. *International Journal of Intelligent Systems and Applications in Engineering*, *12*(1s), 697–703.
[39] Ren, Jie, et al. "A Survey on Recent Advances in the Integration of Discrete Event Systems and Artificial Intelligence." Applied Sciences 16.6 (2026): 3000.
[40] Vo Van An, Trinh Luong Mien, Nguyen Van Binh (2026)." A Comprehensive Survey of UAV Control Algorithms: Integrating Classical Methods with Artificial Intelligence for Enhanced Trajectory Tracking." AIMS Electronics and Electrical Engineering, 10(1), 26–53.
[41] Palazzo, Stefano, Federica Palazzo, and Giovanni Zambetta. "The Role of Artificial Intelligence in Advanced Engineering: Current Trends and Future Prospects." Journal of Intelligent Communication 4.1 (2025): 1–30
[42] Liu, Sicong, et al. "Enabling resource-efficient AIoT system with cross-level optimization: A survey." IEEE Communications Surveys & Tutorials 26.1 (2023): 389–427.
[43] Li, Junfei, and Simon X. Yang. "Intelligent escape of robotic systems: A survey of methodologies, applications, and challenges." Journal of Intelligent & Robotic Systems 109.3 (2023): 55.
[44] Selvaraj, R., Kuthadi, V. M., & Baskar, S. (2023). Smart building energy management and monitoring system based on artificial intelligence in smart city. *Sustainable Energy Technologies and Assessments*, *56*, 103090.
[45] Hernández, J. L., et al. (2024). Challenges and opportunities in European smart buildings energy management: A critical review. *Renewable and Sustainable Energy Reviews*, *199*, 114472.
[46] Agarwal, Y., et al. (2010). Occupancy-driven energy management for smart building automation. In *Proceedings of the 2nd ACM Workshop on Embedded Sensing Systems for Energy-efficiency in Building*.
[47] Jatavallabha, A. (2024). Tesla's autopilot: Ethics and tragedy. *arXiv Preprint arXiv:2409.17380*.
[48] Shutko, J., et al. (2018). *Driver Behavior while Operating Partially Automated Systems: Tesla Autopilot Case Study*. No. 2018-01-0497. SAE Technical Paper.
[49] Gao, J., & Jamidar, R. (2014). Machine learning applications for data center optimization. *Google White Paper*, *21*, 1–13.
[50] DeepMind AI reduces Google data centre cooling bill by 40%. Google DeepMind. Retrieved May 1, 2025, from https://deepmind.google/discover/blog/deepmind-ai-reduces-google-data-centre-cooling-bill-by-40/
[51] Liu, W., et al. (2020). "A method of NC machine tools intelligent monitoring system in smart factories. *Robotics and Computer-integrated Manufacturing*, *61*, 101842.
[52] Factory of the future. What is a smart factory?. Siemens Digital Industries Software. Retrieved May 1, 2025, from https://resources.sw.siemens.com/en-US/white-paper-electronics-semiconductors-smart-manufacturing-what-is-a-smart-factory/
[53] Alemão, D., et al. (2022). "How to design scheduling solutions for smart manufacturing environments using RAMI 4.0?. *IEEE Access*, *10*, 71284–71298.
[54] SMART MANUFACTURING: Building the Factory of Tomorrow. Retrieved May 1, 2025, from https://assets.new.siemens.com/siemens/assets/api/uuid:4259772d-c315-4748-b571-bea6bf983f8c/final-factorytomorrow-wp.pdf
[55] Burra, R., *et al*. (2014). GE Brilliant wind farms. In *2014 IEEE Symposium on Power Electronics and Machines for Wind and Water Applications* (pp. 1–10). Milwaukee, WI, USA. 10.1109/PEMWA.2014.6912227
[56] BETTER FOR YOUR BUSINESS BETTER FOR OUR PLANET. Honeywell.com. Retrieved May 1, 2025, from https://automation.honeywell.com/us/en/campaigns/hts/hvac-systems

Astha Singh, Ashish Bisht*, Deepanshu Tyagi, R. K. Tripathi*

Chapter 11
Route Optimization Techniques in Google Maps Using Graph Theory

Abstract: In this study, Dijkstra's graph theory approach is used to analyze Google Maps' route optimization. This study uses Python implementations and manual validation to show the reliability and effectiveness of Dijkstra's technique by modeling the road networks of Punjab and Uttarakhand as weighted graphs. The findings demonstrated how theoretical graph algorithms closely resemble actual navigation systems when compared to global data sources like Google Maps.

Keywords: Graph theory, route optimization, Dijkstra's algorithm, Python code, weighted graph

11.1 Introduction

In the last few years, navigation technology has grown essential for logistical planning and daily travel. Finding the most effective path between two sites is a basic problem in such systems. By representing the road network as a graph, with cities or intersections serving as vertices and the connected roads acting as edges with weights dependent on trip time or distance, graph theory provides strong tools to address this issue. This study uses the well-known shortest-path algorithm Dijkstra's algorithm to determine the best route between the northernmost and southernmost cities in the Indian states of Uttarakhand and Punjab. The approach was simulated on network models constructed from city-to-city distance data using Python programming. To verify the findings, the outputs were then compared with data from Google Maps and manually estimated distances. The project examines the effectiveness of graph-based methods in resolving actual routing issues through this comparison. The main objective of this study is to demonstrate how theoretical algorithms can be applied to real-world geographic systems. Additionally, it looks at variables like data accuracy, network con-

***Corresponding author: Ashish Bisht,** Department of Mathematics, Graphic Era Hill University, Uttarakhand, India, e-mail: asish10bisht@gmail.com
***Corresponding author: R. K. Tripathi,** Department of Mathematics, Graphic Era Hill University, Uttarakhand, India, e-mail: pvats2009@gmail.com
Astha Singh, Department of Mathematics, Graphic Era Hill University, Uttarakhand, India, e-mail: asthasingh2904@gmail.com
Deepanshu Tyagi, Department of Mathematics, Graphic Era Hill University, Uttarakhand, India, e-mail: deeptyagi73@gmail.com

 | https://doi.org/10.1515/9783112211991-011

nectivity, and road conditions that affect route optimization. In the end, this study shows how graph theory may be used to plan infrastructure in complicated networks and increase transportation efficiency.

This chapter is organized as follows: Section 11.2 presents the literature survey. Section 11.3 describes the workflow of the proposed methodology. Section 11.4 discusses the results as experimental results and comparative analysis. Finally, Section 11.5 concludes the chapter.

11.2 Literature Survey

For a long time, road networks have been represented as graphs using graph theory as a basis for transportation problem resolution. Roads are viewed as edges and cities as nodes in these models, where weights stand in for travel durations or distances. The foundation for contemporary route optimization algorithms was laid by Ceder [1], who highlighted the application of network-based scheduling in public transportation systems.

Basic algorithms like Floyd-Warshall and Dijkstra's, which are frequently used to identify the shortest pathways, were described in detail by Natarajan and Balaji [2]. This study directly applies their theoretical foundation. In order to verify our findings against real-time data from Google Maps, Wi [3] emphasized how changes in edge weights – such as distance, time, or cost – can have a substantial impact on route selection.

The use of graph-based techniques is further supported by real-world applications. In order to optimize Mumbai's bus routes, Kejriwal and Temrikar [6] used Dijkstra's algorithm, demonstrating its efficacy in metropolitan Indian settings. Similar algorithms were used to European metro systems by Błocki and Golędzinowski [4], proving their scalability and effectiveness in dense networks.

While Agustin et al. [7] and Al-Tameemi [8] demonstrated the viability of Python-based simulations and multimodal routing adaptations, Van Steen [5] provided insights into how graph structure shapes algorithm performance. These experiments demonstrate how adaptable the algorithm is at different scales and kinds of transportation.

Furthermore, Gómez Rivera et al. [10] demonstrated how academic models can be used to improve real-world applications by integrating graph algorithms with Google Maps APIs for logistics optimization. Kumar et al. [9] used graph theory to assess the connectivity of Indian state roads, offering a national background that bolsters the present study's emphasis on Punjab and Uttarakhand.

In addition to pointing out issues like data limits, dynamic situations, and infrastructure variability – all of which are considered in our analysis – these studies collectively validate the dependability of Dijkstra's algorithm for route optimization.

11.3 Methodology

The shortest routes between the northernmost and southernmost cities in the Indian states of Punjab and Uttarakhand are found using Dijkstra's algorithm, a well-known graph theory technique. The main goal is to confirm if algorithmic models produce findings that are in line with manual estimations and those provided by Google Maps when applied to real-world data.

The following stages make up the methodology:

11.3.1 Gathering Information and Creating Graphs

We started by determining each state's major cities. The cities that were chosen to be the northern and southernmost were:

1. Punjab: Mansa (South) to Pathankot (North) (shown in Figure 11.1)
2. Uttarakhand: from Udham Singh Nagar in the south to Uttarkashi in the north (shown in Figure 11.2)

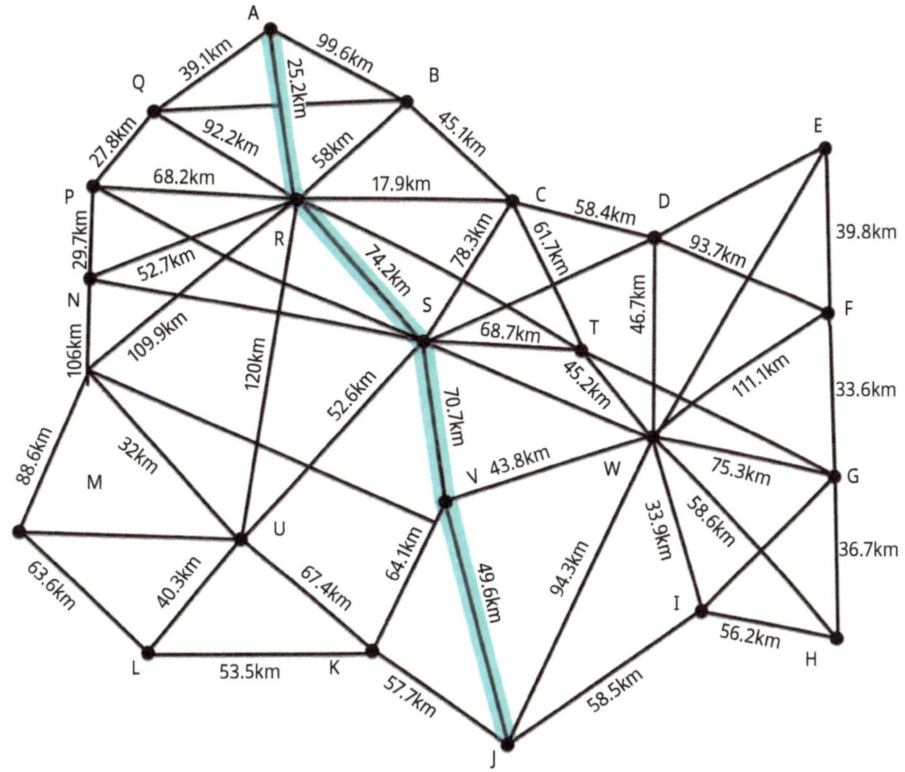

Figure 11.1: Graphical representation for the state of Punjab.

Road edges were weighted based on actual distances (in kilometers) obtained from reliable internet resources such as Google Maps, and cities were considered as nodes in a network. (Table 11.1 shows Punjab cities and Table 11.3 shows cities of Uttarakhand)

The road network in each state was organized as an undirected, weighted graph. The shortest pathways might be calculated using Dijkstra's method.

11.3.2 Estimating Routes by Hand Prior to Employing Dijkstra's Technique

We used the following procedure to manually determine the shortest path: Maps of major cities were created between the northern and southernmost areas. A road atlas or topographic map was used to map the route. The overall travel length was calculated by adding the recorded distances between cities as given in Table 11.2 and Table 11.4. The result of the algorithm was validated using this as the baseline. For example, the manually calculated distance between Uttarkashi and Udham Singh Nagar was 423 miles, and the distance between Pathankot and Mansa was 319.7 km.

Table 11.1: Cities represented as nodes.

Symbol	City
A	Pathankot
B	Hushiyarpur
C	Jalandhar
D	SBS Nagar
E	Rupnagar
F	SAS Nagar
G	Fathegarh Sahib
H	Patiala
I	Sangrur
J	Mansa
K	Bathinda
L	Sri Muktsar Sahib
M	Fazilka
N	Firozpur
O	Tarn Tarn
P	Amritsar
Q	Gurdaspur
R	Kapurthala
S	Moga
T	Ludhiana
U	Faridakot
V	Barnala

11.3.3 Using the Dijkstra's Algorithm

Finding the shortest path between a single source node and every other node in a weighted network is the goal of Dijkstra's algorithm. Python and Java were used to implement it in our study.

Algorithm Execution Steps:

I. Setting Up: A rough distance value was given to each node. Every other node was set to infinity, while the source node (Pathankot or Uttarkashi) was initialized to 0.

II. Priority Queue (Min-Heap): The node with the least estimated distance was repeatedly chosen using a priority queue.

III. Edge Relaxation: The method determined a new estimated distance for every nearby node. The value was updated if it was less than the one that had been previously recorded.

IV. Termination: After identifying the shortest routes between the source and each node in the graph, the algorithm stopped running once every node had been visited (fixed).

11.3.4 Tabular Implementation (Example in Punjab)

A simplified version of Dijkstra's table for Punjab, with A standing for Pathankot and J for Mansa, is shown in Table 11.2:

Through constant refinement of the shortest path estimation, this iterative approach finally matched the manually determined distance of 319.7 km.

11.3.4.1 Punjab Results

- **Manual Calculation: 319.7 km**
- **Python's Dijkstra: 318 km**
- **307 miles on Google Maps**

Road changes and optimizations that are not visible to static graph models are responsible for the small variation in Google Maps distance.

11.3.4.2 Tabular Implementation (Example in Uttarakhand)

Uttarakhand application: The road system in Uttarakhand was subjected to the same process. The route between Uttarkashi and Udham Singh Nagar was charted through the cities of Nainital, Haridwar, and Dehradun as shown in Table 11.4.

Table 11.2: Dijkstra's table for Punjab.

Fixed source	B	Q	R	P	S	C	V	U	K	J
A	∞	∞	∞	∞	∞	∞	∞	∞	∞	∞
A	99.6	39.1	125.2	∞	∞	∞	∞	∞	∞	∞
A, Q	99.6	-	125.2	111.9	∞	∞	∞	∞	∞	∞
A, Q, B	-	-	125.2	111.9	239.6	∞	∞	∞	∞	∞
A, Q, B, R, P	-	-	-	-	117.5	144.7	∞	245.2	∞	∞
A, Q, B, R, P, S	-	-	-	-	-	143.1	∞	245.2	∞	∞
A, Q, B, R, P, S, C	-	-	-	-	-	-	270.1	245.2	279.7	∞
A, Q, B, R, P, S, C, U	-	-	-	-	-	-	270.1	-	279.7	∞
A, Q, B, R, P, S, C, U, V	-	-	-	-	-	-	-	-	279.7	319.7
A, Q, B, R, P, S, C, U, J	-	-	-	-	-	-	-	-	-	319.7

Table 11.3: Cities of Uttarakhand as node.

Symbol	City
Utt	Uttarkashi
De	Dehradun
Te	Tehri garhwal
Ha	Haridwar
Pa	Pauri
Ru	Rudraprayag
Ch	Chamoli
Ba	Bageshwar
Pit	Pithoragarh
Al	Almora
Na	Nanital
Ud	Udham singh Nagar
Cha	Champavat

Manual calculation done for Uttarakhand are as follows:

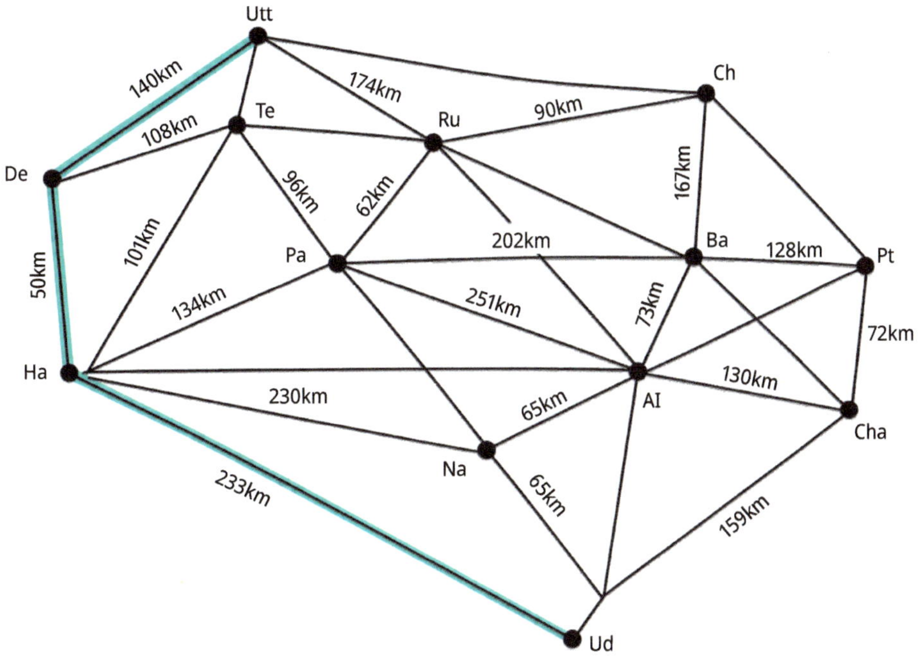

Figure 11.2: Graphical representation for the state of Uttarakhand.

Using Dijkstra's algorithm:

Table 11.4: Dijkstra's table for the state of Uttarakhand.

Fixed Source	D	T	H	P	A	U'
U	∞	∞	∞	∞	∞	∞
U,T	140	105	∞	∞	∞	∞
U,T,P	140	-	206	201	∞	∞
U,T,P,H	140	-	206	-	452	∞
U,T,P,H,D,	140	-	-	-	452	∞
U,T,P,D,U'	-	-	-	-	452	434
U,T,P,D,U',A	-	-	-	-	452	-

Steps in the algorithm initialization of nodes (Uttarkashi = 0, others = ∞).

Route estimation and edge relaxation were performed.

Dijkstra's algorithm was used to calculate the final output demonstrates great agreement with computations made by hand. The output from Google Maps did, however, show a small variance, which is probably the result of real-time route improvements such road widening, construction detours, or dynamic traffic-based rerouting. There are slight variations since Google Maps uses real-time data, such as traffic, tolls, and road closures, while our method employs static, fixed-weight edges.

Implementation using Python. We used Python to implement Dijkstra's algorithm. The actual reasoning employed is demonstrated in the code sample below:

```
import heapq

def dijkstra(graph, start, end):
    pq = [(0, start)]   # (distance, node)
    distances = {node: float('inf') for node in graph}
    distances[start] = 0
    previous_nodes = {node: None for node in graph}

    while pq:
        current_distance, current_node = heapq.heappop(pq)
        if current_node == end:
            break
        if current_distance > distances[current_node]:
            continue
        for neighbor, weight in graph[current_node]:
            distance = current_distance + weight
            if distance < distances[neighbor]:
                distances[neighbor] = distance
                previous_nodes[neighbor] = current_node
                heapq.heappush(pq, (distance, neighbor))

    path = []
    current_node = end
    while current_node is not None:
        path.insert(0, current_node)
        current_node = previous_nodes[current_node]

    return distances[end], path

# Example Graph Representation for Cities (Adjacency List)
# Cities connected with distances (in km)
graph = {
    'Pathankot': [('Hushiyarpur', 99), ('Gurdaspur', 39), ('Kapurthala', 125)],
    'Gurdaspur': [('Pathankot', 39), ('Hushiyarpur', 90), ('Kapurthala', 91), ('Amritsar', 75)],
    'Amritsar':  [('Gurdaspur', 75), ('Kapurthala', 68), ('Tarn tarn', 29)],
    'Jalandhar': [('Hushiyarpur', 45), ('Kapurthala', 29), ('Moga', 78), ('Ludhiana', 61), ('SBS Nagar', 68)],
    'Hushiyarpur': [('Pathankot', 107), ('Kapurthala', 60), ('Gurdaspur', 91), ('Jalandhar', 45), ('Moga', 124)],
    'Tarn tarn': [('Amritsar', 28), ('Kapurthala', 52), ('Firozpur', 93)],
    'Kapurthala': [('Pathankot', 125), ('Hushiyarpur', 60), ('Jalandhar', 29), ('Moga', 74), ('Faridkot', 118), ('Firozpur', 109), ('Tarn ('Amritsar', 68), ('Gurdaspur', 92)],
    'Rupnagar': [('SBS Nagar', 76), ('Malerkotla', 97), ('SAS Nagar', 39)],
    'Firozpur': [('Kapurthala', 109), ('Faridkot', 32), ('Sri Muktsar Sahib', 68), ('Fazilka', 88), ('Tarn tarn', 92)],
    'Moga': [('Kapurthala', 74), ('Jalandhar', 78), ('Ludhiana', 68), ('Barnala', 70), ('Faridkot', 50), ('Firozpur', 55)],
    'Ludhiana': [('Jalandhar', 61), ('SBS Nagar', 92), ('Malerkotla', 46), ('Barnala', 74), ('Moga', 86)],
    'SAS Nagar': [('Rupnagar', 39), ('SBS Nagar', 93), ('Ludhiana', 92), ('Malerkotla', 101), ('Fatehgarh Sahib', 33)],
    'SBS Nagar': [('Jalandhar', 68), ('Ludhiana', 92), ('Malerkotla', 45), ('SAS Nagar', 93), ('Rupnagar', 76), ('Fatehgarh Sahib', 67)],
    'Faridkot': [('Kapurthala', 118), ('Firozpur', 109), ('Sri Muktsar Sahib', 40), ('Bhatinda', 67), ('Barnala', 95), ('Moga', 50)],
    'Fazilka': [('Firozpur', 88), ('Faridkot', 93), ('Sri Muktsar Sahib', 57)],
    'Sri Muktsar Sahib': [('Fazilka', 57), ('Firozpur', 68), ('Faridkot', 40), ('Bhatinda', 53), ('Barnala', 116)],
    'Bhatinda': [('Sri Muktsar Sahib', 53), ('Faridkot', 67), ('Barnala', 64), ('Mansa', 57)],
    'Barnala': [('Moga', 70), ('Ludhiana', 74), ('Malerkotla', 43), ('Sangrur', 94), ('Mansa', 94), ('Bhatinda', 64), ('Faridkot', 95)],
    'Malerkotla': [('Ludhiana', 46), ('SBS Nagar', 45), ('Rupnagar', 97), ('SAS Nagar', 101),
                   ('Fatehgarh Sahib', 65),('Patiala', 58), ('Sangrur', 34), ('Manchur', 34), ('Barnala', 58), ('Mansgor', 94),('Setedrua', 94), ('Barnāha', 43)],
    'Mansa': [('Bhatinda', 57), ('Barnala', 49), ('Malerkotla', 94), ('Sangrur', 60)],
    'Sangrur': [('Mansa', 60), ('Barnala', 39), ('Malerkotla', 34), ('Fatehgarh Sahib', 82), ('Patiala', 56)],
    'Fatehgarh Sahib': [('SAS Nagar', 33), ('SBS Nagar', 67), ('Malerkotla', 45), ('Sangrur', 82), ('Patiala', 36)],
    'Patiala': [('Fatehgarh Sahib', 36), ('Malerkotla', 58), ('Sangrur', 56)]
}

start = 'Pathankot'   # Northern city
end =  'Mansa'        # Southern city

shortest_distance, path = dijkstra(graph, start, end)
print(f"Shortest Distance: {shortest_distance} km")
print("Path: " + " -> ".join(path))
```

Figure 11.3: Python program to implement Dijkstra's algorithm.

11.3.4.3 Uttarakhand Results

The quickest way to get from Uttarkashi to Udham Singh Nagar in Uttarakhand is as follows: Uttarkashi → Tehri → Dehradun → Haridwar → Nainital → Udham Singh Nagar as obtained from Figure 11.2.

Dijkstra in Python: 423 km of output
Calculated by hand: 423 km
Distance using Google Maps: 407.3 km

The algorithm's output in this instance precisely corresponds to the manual computation as shown in Table 11.5. Our modeled graph may have excluded minor roads, bypasses, or newly built highways, which could account for the 15.7 km difference when compared to Google Maps. Additionally, there may be quantifiable discrepancies between theoretical and navigational distances due to natural obstacles in hilly regions, such as switchback roads.

11.4 Output and Validation

Table 11.5: The results and comparison of applying Dijkstra's algorithm to the road networks of Punjab and Uttarakhand.

States	Northernmost City	Southernmost city	Manual Calculation (km)	Using Dijkstra's by python (km)	Global data distance (km)
Punjab	Pathankot	Mansa	319.7km	318km	307km
Uttarakhand	Uttarkashi	Udham Singh Nagar	423km	423km	407.3km

By contrasting the shortest routes determined by Python implementation with manually determined distances and those supplied by Google Maps, the outcomes are assessed. Disparities and discrepancies are also examined considering real-world factors including traffic, route availability, and data completeness.

11.4.1 Results and Discussion

The results of the experiment provide compelling evidence for the application of Dijkstra's algorithm to state-level route optimization. Given the constraints of static graph models, the variations from Google Maps, which range from 5 to 15 km, are rather small and to be expected. This research does not imitate the live data, temporary diversions, road conditions, and multiple route algorithms that are a part of real-world routing systems.

However, Dijkstra's algorithm's ability to model geographically realistic road networks and generate results that are almost accurate indicates that graph-based methods can greatly support offline navigation systems, instructional simulations, and transportation planning.

11.4.2 Limitations of Dijkstra's Algorithm

Despite being a useful algorithm for determining the shortest path it work with static weights and Google Maps highly depends upon dynamic weights as it need to update the weight of every edge in couple of seconds, minutes, and hours to provide accurate distance to the user.

It is also time-consuming as it searches for every path connected to the route whether it will take us to the destination or not therefore wasting user's time and resource (fuel and electricity).

11.4.3 Limitations Overcome by A* Algorithm

A* (A star) algorithm is nothing new but the updated version of Dijkstra's algorithm. Now it uses heuristic (guess) approach for better navigation. It smartly guesses which path will take us to the destination rather than exploring another route.

11.4.3.1 Limitations for Hilly Terrain

The algorithm only works well when data is accurate, such as in plains. But in hilly terrain, the map is outdated or incomplete, which leads to unsafe paths or paths with high chances of landslides as satellite view does not guarantee the safety and steepness of road and nearby location. Who knows if the road is so steep, making it impossible for the vehicle to go further. So even the mathematical shortest path may be dangerous.

11.5 Conclusion and Future Work

Graph theory was used in this project to optimize the route between Uttarakhand and Punjab, with the help of Dijkstra's algorithm both manually and with Python code. Using Dijkstra's algorithm to actual geographic data showed that it was a useful tool for resolving the shortest path problem, yielding accurate and dependable answers.

Despite advanced algorithms of graph theory, misleading routes remain common in the hilly terrains of Uttarakhand due to the lack of proper information and technological advancement. These challenges can be overcome by smart approaches of graph theory, government, and local administrative support.

11.5.1 Future Works to Reduce the Limitations

Using graph theory, Google Maps can improve these limitations and hence providing better experience for the user.

Calculating the distance between two nodes with paved or unpaved road with steepness (high or low, if high slope then considering the edge weight high and if low steep then considering the edge as low weight) with changes of landslide or not. In short:

Weight of edge = distance + paved/unpaved + slope + landslide zone

Providing more than one path for the destination depending upon the risk and mode of transportation:

"Fastest + high risk route", "long route + safe", "balanced route"

Feedback program by the tourist and labeling the road as per their quality:

Poor road = increase edge weight

Good road = decrease edge weight

Landslide/roadblock = block edge temporary

At night, the map should be updated to avoid leading tourist into unsafe places like forests or abandoned areas.

References

[1] Toman, J., & Olszewska, J. I. (2014, November). Algorithm for graph building based on Google Maps and Google Earth. In *2014 IEEE 15th International Symposium on Computational Intelligence and Informatics (CINTI)* (pp. 55–60). IEEE.

[2] Natarajan, B. L., & Balaji, M. K. (2019). Application of graph theory in online network services to determine the shortest journey. *International Journal of Advanced Networking and Applications, 10*(5). 4030–4034.

[3] Wi, Y. N. *Graph Theory: Application of Graphs in Determining the Most Efficient Highway Routes.*

[4] Błocki, W., & Golędzinowski, W. (2024). An application of the graph theory to the analysis of metro networks in selected European cities. *The Book of Articles*, 15–26.

[5] Van Steen, M. (2010). *Graph Theory and Complex Networks. An Introduction* (vol. 144, 1).

[6] Kejriwal, A., & Temrikar, A. (2019). Graph theory and Dijkstra's algorithm: A solution for Mumbai's BEST buses. *The International Journal of Engineering and Science (IJES), 8*(10). 40–47.

[7] Agustin, N. D., Adawiyah, R., & Makhfudloh, I. I. *Implementation of Graph Applications (Dijkstra's Algorithm) and PYTHON in Determining the Shortest Path from Boarding House to Campus.*

[8] Al-Tameemi, H. L. H. (2014). Using dijkstra algorithm in calculating alternative shortest paths for public transportation with transfers and walking (Doctoral dissertation).

[09] Kumar, J. S., Archana, B., Muralidharan, K., & Kumar, V. S. (2025). Graph Theory: Modelling and Analyzing Complex System. *Metallurgical and Materials Engineering*, *31*(3). 70–77.

[10] Gómez Rivera, U. Á., Pérez Olguín, I. J. C., Pérez Domínguez, L. A., Rodríguez-Picón, L. A., & Méndez-González, L. C. (2023). Distribution route optimization using Floyd-Warshall weighted graph analysis algorithm with Google Maps integration in industry 4.0 context. In *Innovation and Competitiveness in Industry 4.0 Based on Intelligent Systems* (pp. 287–305). Cham: Springer International Publishing.

[11] Wikipedia (for images)

[12] Google maps (for official data of the distances and time taken)

Index

 | https://doi.org/10.1515/9783112211991-012

De Gruyter Series on the Applications of Mathematics in Engineering and Information Sciences

Already Published in the Series

Volume 26
Fuzzy Decision-Making Algorithms: Mathematical Optimization, Metaheuristic Optimization, Supply Chain Management
Kamal Kumar, Shahid Ahmad Bhat, Paul Augustine Ejegwa, Reeta Bhardwaj (Eds.)
ISBN 978-3-11-914266-3, e-ISBN (PDF) 978-3-11-222224-9,
e-ISBN (EPUB) 978-3-11-222254-6

Volume 25
Artificial Intelligence in Cancer: Detection, Diagnosis, and Treatment
Rishabha Malviya, Shivam Rajput, Mukesh Roy, Sathvik Belagodu Sridhar, Sabin Thomas
ISBN 978-3-11-167037-9, e-ISBN (PDF) 978-3-11-167074-4,
e-ISBN (EPUB) 978-3-11-167151-2

Volume 24
Complex Engineering Systems – Modeling and Optimization: Multi-Objective Analysis and Smart Manufacturing Applications
Satyvir Singh, Mukesh Kumar Awasthi (Eds.)
ISBN 978-3-11-172340-2, e-ISBN (PDF) 978-3-11-172346-4,
e-ISBN (EPUB) 978-3-11-172359-4

Volume 23
Fuzzy Clustering Algorithms: Symbiotic Relationship Between Mathematics and Similarity Measures
Rakesh Kumar, Geeta Arora
ISBN 978-3-11-149238-4, e-ISBN (PDF) 978-3-11-149427-2,
e-ISBN (EPUB) 978-3-11-149576-7

Volume 22
Metaverse and Digital Twins: Blockchain and Healthcare Technology
Bhupinder Singh, Rishabha Malviya, Christian Kaunert (Eds.)
ISBN 978-3-11-163629-0,
e-ISBN (PDF) 978-3-11-164689-3,
e-ISBN (EPUB) 978-3-11-164464-6

Volume 21
Reengineering Cyber Security Process: Quality Management
Mehrdad Sepehri Sharbaf
ISBN 978-3-11-144382-9, e-ISBN (PDF) 978-3-11-144463-5,
e-ISBN (EPUB) 978-3-11-144480-2

Volume 20
Digital Blockchain: Big Data, Artificial Intelligence, and Virtual Reality in Healthcare Ecosystem
Rishabha Malviya, Shristy Verma, Sonali Sundram, Harshil Shah
ISBN 978-3-11-157386-1, e-ISBN (PDF) 978-3-11-157428-8,
e-ISBN (EPUB) 978-3-11-157485-1

Volume 19
Artificial Intelligence for Healthcare: Machine Learning and Diagnostics
Rishabha Malviya, Shivam Rajput, Mukesh Roy, Irfan Ahmad, Saurabh Srivastava
ISBN 978-3-11-154758-9, e-ISBN (PDF) 978-3-11-154972-9,
e-ISBN (EPUB) 978-3-11-154984-2

Volume 18
Operational Perspective of Modeling System Reliability: Research Tools for System Dynamics
Deepti Aggrawal, Adarsh Anand, Yoshinobu Tamura, Mohini Agarwal (Eds.)
ISBN 978-3-11-147540-0, e-ISBN (PDF) 978-3-11-147610-0,
e-ISBN (EPUB) 978-3-11-147713-8

Volume 17
Algorithms: Big Data, Optimization Techniques, Cyber Security
Sushil C. Dimri, Abhay Saxena, Bhuvan Unhelkar, Akshay Kumar (Eds.)
ISBN 978-3-11-122800-6, e-ISBN (PDF) 978-3-11-122915-7,
e-ISBN (EPUB) 978-3-11-122963-8

Volume 16
Distributed Transfer Function Method: One-Dimensional Problems in Engineering
Bingen Yang, Kyoungrae Noh
ISBN 978-3-11-075854-2, e-ISBN (PDF) 978-3-11-075893-1,
e-ISBN (EPUB) 978-3-11-075900-6

Volume 15
Machine Learning for Cyber Security
Malik Preeti, Nautiyal Lata, Ram Mangey (Eds.)
ISBN 978-3-11-076673-8, e-ISBN (PDF) 978-3-11-076674-5,
e-ISBN (EPUB) 978-3-11-076676-9

Volume 14
Multiple Criteria Decision-Making Methods: Applications for Managerial Discretion
Mohini Agarwal, Adarsh Anand, Deepti Aggrawal
ISBN 978-3-11-074356-2, e-ISBN (PDF) 978-3-11-074363-0,
e-ISBN (EPUB) 978-3-11-074374-6

Volume 13
Integral Transforms and Applications
Nita H. Shah, Monika K. Naik
ISBN 978-3-11-079282-9, e-ISBN (PDF) 978-3-11-079285-0,
e-ISBN (EPUB) 978-3-11-079292-8

Volume 12
Noise Filtering for Big Data Analytics
Souvik Bhattacharyya, Koushik Ghosh (Eds.)
ISBN 978-3-11-069709-4, e-ISBN (PDF) 978-3-11-069721-6,
e-ISBN (EPUB) 978-3-11-069726-1

Volume 11
Artificial Intelligence for Signal Processing and Wireless Communication
Abhinav Sharma, Arpit Jain, Ashwini Kumar Arya, Mangey Ram (Eds.)
ISBN 978-3-11-073882-7, e-ISBN (PDF) 978-3-11-073465-2,
e-ISBN (EPUB) 978-3-11-073472-0

Volume 10
Meta-heuristic Optimization Techniques: Applications in Engineering
Anuj Kumar, Sangeeta Pant, Mangey Ram, Om Yadav (Eds.)
ISBN 978-3-11-071617-7, e-ISBN (PDF) 978-3-11-071621-4,
e-ISBN (EPUB) 978-3-11-071625-2

Volume 9
Linear Integer Programming: Theory, Applications, Recent Developments
Elias Munapo, Santosh Kumar
ISBN 978-3-11-070292-7, e-ISBN (PDF) 978-3-11-070302-3
e-ISBN (EPUB) 978-3-11-070311-5

Volume 8
Mathematics for Reliability Engineering: Modern Concepts and Applications
Mangey Ram, Liudong Xing (Eds.)
ISBN 978-3-11-072556-8, e-ISBN (PDF) 978-3-11-072563-6
e-ISBN (EPUB) 978-3-11-072559-9

Volume 7
Mathematical Fluid Mechanics: Advances on Convection Instabilities and Incompressible Fluid Flow
B. Mahanthesh (Ed.)
ISBN 978-3-11-069603-5, e-ISBN (PDF) 978-3-11-069608-0
e-ISBN (EPUB) 978-3-11-069612-7

Volume 6
Distributed Denial of Service Attacks: Concepts, Mathematical and Cryptographic Solutions
Rajeev Singh, Mangey Ram (Eds.)
ISBN 978-3-11-061675-0, e-ISBN (PDF) 978-3-11-061975-1
e-ISBN (EPUB) 978-3-11-061985-0

Volume 5
Systems Reliability Engineering: Modeling and Performance Improvement
Amit Kumar, Mangey Ram (Eds.)
ISBN 978-3-11-060454-2, e-ISBN (PDF) 978-3-11-061737-5
e-ISBN (EPUB) 978-3-11-061754-2

Volume 4
Systems Performance Modeling
Adarsh Anand, Mangey Ram (Eds.)
ISBN 978-3-11-060450-4, e-ISBN (PDF) 978-3-11-061905-8
e-ISBN (EPUB) 978-3-11-060763-5

Volume 3
Computational Intelligence: Theoretical Advances and Advanced Applications
Dinesh C. S. Bisht, Mangey Ram (Eds.)
ISBN 978-3-11-065524-7, e-ISBN (PDF) 978-3-11-067135-3
e-ISBN (EPUB) 978-3-11-066833-9

Volume 2
Supply Chain Sustainability: Modeling and Innovative Research Frameworks
Sachin Kumar Mangla, Mangey Ram (Eds.)
ISBN 978-3-11-062556-1, e-ISBN (PDF) 978-3-11-062859-3,
e-ISBN (EPUB) 978-3-11-062568-4

Volume 1
Soft Computing: Techniques in Engineering Sciences
Mangey Ram, Suraj B. Singh (Eds.)
ISBN 978-3-11-062560-8, e-ISBN (PDF) 978-3-11-062861-6,
e-ISBN (EPUB) 978-3-11-062571-4